# Arbitration Practice
# in Construction Contracts

*Also of interest*

**The Arbitration Act 1996**
A Commentary
Bruce Harris, Rowan Planterose
and Jonathan Tecks
Published in conjunction with
The Chartered Institute of Arbitrators
0-632-04131-5

# Arbitration Practice

## in Construction Contracts

### Fifth Edition

**Douglas A. Stephenson**
*BSc, CEng, FICE, FIStructE, FCIArb, MConsE*

Foreword by
The Rt. Hon. The Lord Saville of Newdigate

**Blackwell
Science**

© 1982, 1987 Douglas Stephenson and International Thompson Publishing; 1993, 1998, 2001 Douglas Stephenson

Blackwell Science Ltd
Editorial Offices:
Osney Mead, Oxford OX2 0EL
25 John Street, London WC1N 2BL
23 Ainslie Place, Edinburgh EH3 6AJ
350 Main Street, Malden
  MA 02148 5018, USA
54 University Street, Carlton
  Victoria 3053, Australia
10, rue Casimir Delavigne
  75006 Paris, France

Other Editorial Offices:

Blackwell Wissenschafts-Verlag GmbH
Kurfüstendamm 57
10707 Berlin, Germany

Blackwell Science KK
MG Kodenmacho Building
7-10 Kodenmacho Nihombashi
Chuo-ku, Tokyo 104, Japan

Iowa State University Press
A Blackwell Science Company
2121 S. State Avenue
Ames, Iowa 50014-8300, USA

First published 1982
Second edition 1987
Third edition published 1993 by E & FN Spon
Fourth edition published 1998 by
Blackwell Science Ltd
Fifth edition published 2001 by
Blackwell Science Ltd

Set in 10.5/12.5 pt Palatino
by DP Photosetting, Aylesbury, Bucks

DISTRIBUTORS

Marston Book Services Ltd
PO Box 269
Abingdon
Oxon OX14 4YN
(*Orders:*  Tel: 01235 465500
            Fax: 01235 465555)

USA
Blackwell Science, Inc.
Commerce Place
350 Main Street
Malden, MA 02148 5018
(*Orders:*  Tel: 800 759 6102
                 781 388 8250
            Fax: 781 388 8255)

Canada
Login Brothers Book Company
324 Saulteaux Crescent
Winnipeg, Manitoba R3J 3T2
(*Orders:*  Tel: 204 837-3987
            Fax: 204 837-3116)

Australia
Blackwell Science Pty Ltd
54 University Street
Carlton, Victoria 3053
(*Orders:*  Tel: 03 9347 0300
            Fax: 03 9347 5001)

A catalogue record for this title is available from the British Library

ISBN 0-632-05741-6

Library of Congress
Cataloging-in-Publication Data
Stephenson, Douglas A.
    Arbitration practice in construction contracts/ Douglas A. Stephenson; foreword by the Lord Saville of Newdigate. — 5th ed.
        p.  cm.
    Includes bibliographical references and index.
    ISBN 0-632-05741-6 (alk. paper)
    1. Construction contracts — Great Britain.
    2. Civil engineering contracts — Great Britain.
    3. Arbitration and award — Great Britain.   I. Title.

KD1641.S73  2001
343.73'078624 — dc21

                                    00-049819

For further information on
Blackwell Science, visit our website:
www.blackwell-science.com

# CONTENTS

Contents

# Contents

# Contents

# FOREWORD

by The Rt. Hon. The Lord Saville of Newdigate

The Fifth Edition of this book provides invaluable guidance and information to all concerned with the resolution of disputes arising out of construction contracts.

The Arbitration Act 1996 was an attempt to set out our basic law of arbitration in a logical and easily read form, in the hope that by doing so this form of dispute resolution would be improved and promoted, both domestically and internationally. We also made a number of changes to the law as it previously stood. These changes and the reasons for them are discussed in two Reports which we wrote on the Bill and the Act which it became, in February 1996 and January 1997. Copies of these Reports are available from the Department of Trade and Industry.

As those who read the Act will see, under Section 33 the arbitral tribunal has the duty to adopt procedures suitable to the circumstances of the particular case, avoiding unnecessary delay or expense, so as to provide a fair means for the resolution of the matters falling to be determined. The arbitral tribunal will not be able properly to perform this duty unless it has a firm working grasp of what being an arbitrator entails. This book helps greatly in this regard.

In this connection I would like to say a few words on comments I have read in some of the construction arbitration journals. These have criticised the Arbitration Act in a number of respects. As might perhaps be expected, I do not accept these criticisms. I would like to state shortly why I take this view. It is not because I regard the Arbitration Act as the best thing since sliced bread; but it is because I think that those who have made the criticisms to which I shall refer have not properly understood the Act; and seem also not to have considered the two Reports to which I have referred.

The first criticism is directed at the fact that the distinction between domestic and international arbitrations has not been maintained, since the sections that would have done so have not been brought into effect. The immediate reason for this was that it was our view that the sections were discriminatory against non-

English EC nationals and therefore contrary to European law. This view was supported by a Court of Appeal decision in 1996, in respect of which the House of Lords refused leave to appeal. Those who dislike the abandonment of the domestic rules, which gave the court a discretion whether to stay legal proceedings brought in disregard of an arbitration clause, have signally failed to put forward any arguments which would justify the distinction between domestic and other arbitrations as a matter of European law, by which we are bound. This was notwithstanding the fact that we inserted the sections in the Act in order to give those concerned an opportunity to persuade us that the distinction could properly be maintained; an opportunity which was not taken.

There is, however, another reason why I reject this criticism. An arbitration agreement is a contract, usually attached to a commercial bargain of some sort. English commercial law is built on the principle that save in the most exceptional circumstances (such as frustration) parties are to be held to their bargain. Why then should an agreement to arbitrate rather than litigate be treated differently? Why should the Court have the power, with hindsight, to put that agreement on one side if it seems more convenient in the circumstances for there to be litigation rather than arbitration? Of course consumers should be protected and so they are by the provisions applying the European Directives on consumer protection.

It is said that the 1996 Act will mean that parties will cease to use arbitration clauses since there will no longer be the right to try to persuade the court that it would be better to have litigation. I am afraid to say that to my mind such a criticism smacks of muddled thinking. The parties have the right to make what arbitration agreement they like. If they want to, they can have no arbitration agreement, or have an optional arbitration agreement, with the choice of going to court instead. All we have done is to say that if you do make an agreement to arbitrate rather than litigate, you will be held to that bargain unless the arbitration agreement proves to be void or incapable of being performed. In other words your arbitration agreement will mean what it says.

It is also said that the inability of the court to allow litigation instead of arbitration will cause great difficulties in multi-party cases, since instead of one hearing there will have to be a number of arbitrations, which might produce inconsistent results. Again, however, the remedy lies in the hands of the parties and those responsible for drafting the standard forms of contract, for section 35 of the Act allows (if the parties agree) for consolidation and the like. Agreement is vital; for why should a small subcontracting party, who planned for a small arbitration with his contractor if things went wrong, be dragged against his will and at the whim of a court years later, into some blockbuster piece of litigation involving many parties and great time and expense?

Some have bemoaned the fact that the court no longer has power to give summary judgment in cases where there is an arbitration clause but where the Court considers that there is no bona fide defence to the claim, on the grounds that this will mean that unanswerable claims will be delayed while the claimant has to go to arbitration. I fundamentally disagree.

In the first place, to allow the matter to be decided by the court is again to disregard the bargain that the parties have made, that their disputes will be arbitrated, not litigated. 'Ah', say the critics, 'but this is not so, since if there is no bona fide defence, then there is no dispute or difference within the meaning of the arbitration clause!'

That cannot be right. A moment's thought would reveal that if it were right, then the arbitrators would only have jurisdiction to decide cases where there was a bona fide defence, for if there was not, then the arbitration clause would not cover the matter. What is even more absurd is the fact that a losing respondent would be able to challenge an award on the grounds that he never had a bona fide defence, so that the arbitrators lacked jurisdiction to make an award against him.

In the second place, there is no good reason why an arbitral tribunal, exercising the powers given to it by the Act, should not be able to deal with such cases quicker than any court. If in truth there is no defence, arbitrators acting in accordance with their section 33 duties should be able to issue an award more or less at once.

What has also given rise to some criticism is the interrelationship between arbitration and adjudication. I am the first to accept that this is a very important topic indeed, but to my mind criticism of the Arbitration Act is misplaced. This Act can work well and effectively with adjudication, if the latter is properly considered and worked out. Great efforts are being made to develop adjudication and I am hopeful that in the future these two quite different concepts will be able to work properly together.

Finally, there are those who consider that we were wrong to give the arbitrators powers to fit the procedures of an arbitration to the particular dispute to be determined. There are those who cling to the theory, particularly in construction circles, that there is really only one way to run a construction dispute of any length. I believe them to be wrong as, I am glad to say, does Margaret Rutherford QC, who in an article called 'The demise of carbon-copy litigation' in an edition of the *Practical Arbitration Journal*, welcomed the fact that extensive pleadings, discovery, interrogatories, the interminable oral examination and cross examination of witnesses and experts, and the reading of endless documents to the tribunal, should no longer be regarded as the best and only way to conduct these arbitrations; at huge expense to the parties and often taking longer than it took to build the works in question.

In my view the great advantage of arbitration over litigation is that in the former it is very much easier to tailor the procedure to the case in hand, so as to avoid unnecessary delay and expense. No doubt sometimes there will be a need for pleadings, full discovery etc. The Act does not prohibit any particular method of conducting the arbitration, provided that the method fulfils the first principle set out in section 1 of the Act, namely that the object of arbitration is to obtain the fair resolution of disputes by an impartial tribunal without unnecessary delay or expense. It is the slavish adherence to old fashioned, expensive and time-consuming procedures whatever the circumstances which in recent years has done much to make arbitration unpopular among those who pay. The Act is designed to try to correct this state of affairs; and to show those concerned that used properly it is an excellent method of alternative dispute resolution. This book has the same object in view and I commend it to all concerned with arbitration.

*Mark Saville*

# PREFACE TO THE FIFTH EDITION

The first edition of this book was published in 1982 under the somewhat restrictive title *Arbitration for Contractors*. It was however recognised, even at that early stage, that the book should not be concerned solely with arbitration from the blinkered viewpoint of one of the parties – usually the claimant – and that it was necessary to cover arbitration in general, with particular reference to construction contract disputes. The second, third and fourth editions appeared in 1987, 1993 and 1998 respectively, the third and fourth editions under the corrected title. Revisions were needed to take account of numerous developments in both statute and case law, of which by far the most important was the enactment of the Arbitration Act 1996 (reproduced herein as Appendix B).

The pace of change in both statute and case law affecting contractual disputes has not however slackened since 1998, when the fourth edition appeared. Statutory changes have generally been consistent with the objective defined in section 1 of the 1996 Act: 'to obtain the fair resolution of disputes by an impartial tribunal without unnecessary delay or expense'. That aim has been pursued in the Civil Procedure Rules 1998 and in Part II of the Housing Grants, Construction and Regeneration Act 1996, which introduced adjudication as a statutory requirement and which also came into force, together with the associated Scheme for Construction Contracts, in 1998. These innovations have resulted in a need for changes in standard forms of construction contract, most of which have been revised to make provision for adjudication and for other statutory developments. In addition the Construction Industry Model Arbitration Rules (CIMAR) were published, in 1998, by the Society of Construction Arbitrators, and have been recognised by the two major construction institutions. A commentary on the rules appears in Chapter 2. Provisions for *the avoidance* and settlement of disputes have become standard features of many of the standard forms, and are also outlined in Chapter 2.

The more general subject of dispute avoidance and management, covering measures designed to reduce the likelihood of disputes arising, and procedures to be followed when they do arise, are dealt with in Chapter 11 of the book.

Much case law has been generated during the past few years by statutory innovations and in particular by the adjudication provisions of the Construction Act and the Scheme associated with it. The more important precedents in this category are summarised in Chapter 11. Another important precedent was created in 1998, when the House of Lords, in *Beaufort Developments (NI)* v. *Gilbert Ash NI*, overturned a judgment by the Court of Appeal in *Northern Regional Health Authority* v. *Derek Crouch Construction* (1984). This judgment and its consequences are dealt with in Chapter 4.

During the 18 years that have elapsed since publication of the first edition the objective of this book has remained unchanged: namely to explain, in jargon-free language, the use of arbitration as a means of resolving contractual disputes and in particular those arising from construction contracts. Happily the use of simple language has in recent years also become a feature of the various enactments referred to in the book, obviating any need to translate from jargon into plain English.

The Arbitration Act 1996 (reproduced herein as Appendix B) has now been in existence for long enough to confirm its excellence and to generate a body of case law that has demonstrated the wisdom of those by whom it was drafted. It provides a clear, concise and logical statement of the English law of arbitration, replacing the Act of 1950 and the numerous amendments contained in other enactments under a variety of titles, some of which did not expressly refer to arbitration.

Such substantive changes as were introduced in the 1996 Act are aimed especially at the principles of fairness and party autonomy; at eliminating the power of the Court to disregard valid arbitration agreements, and at strengthening the arbitrator's powers to conduct references justly, expeditiously and economically. The arbitrator's power to award only simple interest was widened in the 1996 Act to an express power under section 49, to award simple or compound interest; recognising the reality of commercial transactions, increasing the incentive to the prompt settlement of debts, and setting an example which could sensibly be followed in litigation. Intervention by the courts has been discouraged by strengthening the arbitrator's powers to deal with situations that previously required the court's jurisdiction: and in particular to deal with situations arising from the reluctance of a party – usually the respondent – to comply with the arbitrator's directions. Construction arbitrators and others involved in the resolution of construction disputes should study the 1996 Act and use the increased powers it provides to enhance the reputation of arbitration as the preferred forum for the resolution of such disputes.

I am delighted and honoured that Lord Saville, who succeeded Lord Mustill and Lord Steyn as Chairman of the Departmental Advisory Committee on Arbitration Law responsible for drafting the 1996 Act, and was in the Chair during the period when the final draft Bill was prepared, has written the Foreword to this book. In doing so Lord Saville has taken the opportunity to reply to some of the criticisms that were expressed of the 1996 Act, and to explain clearly and concisely the reasoning underlying the Act's provisions. Readers, and especially those who practice as arbitrators, are urged to take note of Lord Saville's words, and to apply them robustly in order to achieve the objectives so clearly set out in section 1 of the 1996 Act. If this book assists them in doing so it will have served its purpose.

In preparing my commentaries in Chapter 2 on the many standard forms of construction contract I have received helpful advice on present and future developments from several of my friends and colleagues; notably Professor John Uff QC, Geoffrey Hawker, Arthur Appleton, Tracey Donaldson and John Pape, whose assistance is gratefully acknowledged.

I have used the male pronoun throughout when referring to the arbitrator and to others involved in the arbitration process: not because they are necessarily male, but because alternatives sometimes adopted are clumsy and in my view unnecessary. Readers are asked to accept my assurance that in law, as in life, the male embraces the female.

Finally I wish to express my thanks to Julia Burden and her staff for their help and encouragement throughout the gestation stages of this edition, and for the efficiency with which they have dealt with the production stages.

*Douglas A. Stephenson*
*October 2000*

# CHAPTER ONE
# INTRODUCTION

## 1.1   Synopsis

The basic principle of arbitration, namely the determination of a dispute arising from a contract between two parties by a third party chosen by the contracting parties, is especially relevant where, as in construction contracts, many of the issues are of a technical nature. Arbitration depends for its efficacy upon a framework of law within which it is recognised: for without such a framework the award of the arbitrator could be worthless if the losing party chose to ignore it. Enforcement of arbitration awards is available through the courts of England and of most other civilised countries: but English law, with which this book is concerned, requires that the arbitration proceedings must have been conducted, and the award made, in accordance with the law.

## 1.2   Definition

'An arbitrator is a private extraordinary judge between party and party, chosen by their mutual consent to determine controversies between them. And arbitrators are so called because they have an arbitrary power: for if they observe the submission and keep within due bounds, their sentences are definite from which there lies no appeal.'

The words of Raymond LCJ, expressed in the eighteenth century, still provide a valid definition: for if the phrase 'due bounds' is taken to mean 'the law,' there is indeed no appeal from an arbitrator's award. The limited grounds upon which an appeal may lie cover only issues of law and failure by the arbitrator to comply with the rules of natural justice: that is, to act fairly (see Chapter 10).

A more recent, and more succinct, definition of arbitration is given in the *Shorter Oxford English Dictionary*:

'The settlement of a matter at issue by one to whom the parties agree to refer their claims in order to obtain an equitable decision.'

Arbitration is a voluntary procedure, available as an alternative to litigation, and enforceable through the courts where the parties have entered into a valid arbitration agreement. In such cases the right of either party to have the dispute resolved by arbitration will be upheld by the court under section 9 of the Arbitration Act 1996 (subsequently referred to as the 1996 Act: see Appendix B), under which the court has a duty to grant a stay to an action brought in contravention of an arbitration agreement.

## 1.3 Legal framework

This book is concerned with arbitration under English law which applies, under section 2 of the 1996 Act, in England, Wales and Northern Ireland, but not in Scotland; and to any arbitration commenced on or after the date on which the main provisions of the 1996 Act came into force, namely 31 January 1997, notwithstanding that the arbitration agreement from which it arises may have been made before that date. In the case of arbitrations commenced before 31 January 1997 the relevant statutes are the Arbitration Acts of 1950, 1975 and 1979, which Acts are subject to many amendments. These acts are applicable in England and Wales but not in Northern Ireland (where the relevant statute is the Arbitration Act (Northern Ireland) 1937), or in Scotland.

English law of arbitration has, however, a far wider influence than the above outline implies. It has been adopted as the basis for arbitration law in many of the former British colonies and protectorates; and it is often adopted as the contractual procedure for dispute resolution in international contracts, especially where, as is often the case, the language of the contract is English. The choice of English law of arbitration as governing dispute resolution procedure under an international contract does not necessarily govern the choice of the proper law of the contract, although there is a presumption to that effect where the contract is silent on such matters.

There are good reasons why English law of arbitration should be adopted in international contracts, even when the proper law of the contract is not English law. First, English commercial law is more highly developed and sophisticated than in any other legal system.

Second, it forms the basis of many other legal systems throughout the world, and is therefore more readily accepted and understood than other systems. And third, the general use of the English language in international contracts reduces the problems of interpretation where English law of arbitration is adopted.

## 1.4 *English law of arbitration*

Arbitration in England is known to have been recognised in common law since the beginning of the seventeenth century: indeed the system is claimed to be as old as legal history. The first statute governing arbitration was the Arbitration Act 1697; and since that date there have been many re-enactments. In recent times one of the most important re-enactments was the Arbitration Act 1950, which codified earlier common and statute law, and remained substantially unchanged, apart from a change in appeal procedure, until the 1996 Act became law. The 1950 Act had been the subject of one major and many minor amendments, introduced by way of the Arbitration Acts of 1975 and of 1979, the Supreme Court Act 1981 and the Administration of Justice Act 1982; with the result that the legislation was criticised as being fragmented and not readily accessible, especially to lay disputants and arbitrators.

That was the situation in existence towards the end of the 1980s, when a question arose as to whether or not the United Nations Commission on International Trade Law (UNCITRAL) Model Law on International Commercial Arbitration should be adopted as the basis of English arbitration law. The Departmental Advisory Committee on Arbitration Law under the chairmanship of Mustill LJ (now Lord Mustill) in its Report of June 1989 advised against so doing, but recommended instead that there should be a new and improved arbitration act for England, Wales and Northern Ireland.

The objectives of the new legislation had been clearly defined, and included such aims as being a statutory enactment of the principles of both existing statute and common law, being set out in a logical order and expressed in clear language comprehensible to the layman, being applicable to domestic and international arbitrations alike, and as having the same structure as the UNCITRAL Model Law. The objectives were however more easily stated than achieved: for it was not until February 1996, after several drafts had been prepared and considered, that the DAC under its later chairman Saville LJ (now Lord Saville) produced its Report and Arbitration Bill, which met with the almost universal acclaim of

3

practitioners in, and users of, arbitration. The Bill's passage through parliament was rapid, and Royal Assent was given on 17 June 1996.

Although the intention in drafting the Bill had not included the introduction of substantial changes in arbitration law, a number of relatively minor, but valuable, changes have been made. The 1996 Act, unusually, commences with a statement of general principles; section 1 defining the object of arbitration as being

'to obtain the fair resolution of disputes by an impartial tribunal without unnecessary delay or expense; [that] the parties should be free to agree how their disputes are resolved, subject only to such safeguards as are necessary in the public interest; [and that] ... the court should not intervene except as provided by this Part.'

The part referred to is Part I of the Act, which for practical purposes is the whole of the Act.

The arbitration tribunal (referred to as 'the arbitrator' in this book) has an express duty, under section 33 of the 1996 Act, to proceed 'avoiding unnecessary delay or expense'; and the arbitrator is expressly empowered to rule on his own jurisdiction (subject of course to safeguards); to award simple or compound interest; to dismiss a claim for want of prosecution; and to order a claimant to provide security for costs of the arbitration. Emphasis is laid on fairness and on party autonomy. About half of the 84 sections of Part I of the 1996 Act begin with the words: 'unless otherwise agreed by the parties' (or words to similar effect) and they are followed by fall-back provisions applicable in the absence of agreement. Agreement in this context means agreement in writing; and that word is defined in section 5 of the 1996 Act to include agreements in an exchange of correspondence, and an agreement evidenced in writing.

Another innovation is the immunity from suit given both to arbitrators and to appointing authorities under section 29 and section 74 respectively of the 1996 Act, reflecting the immunity enjoyed by the judiciary. In the fourth edition of this book reservations were expressed as to the wisdom of this change, which the author thought might result in a lowering of the standards of competence of arbitrators. In the event there appears to have been no such deterioration since the Act came into force: its beneficial effects resulting from its clear drafting and its widening of the powers of arbitrators to dispense justice appears to have led to a general improvement in standards of performance that has more than compensated for any possibly adverse effects of the immunities given.

# Introduction

It is perhaps strange that within four years of the introduction of immunity for arbitrators and for appointing authorities the immunity enjoyed for more than 30 years by advocates in both civil and criminal actions in the courts should have been withdrawn by the House of Lords. In *Arthur J.S. Hall & Co* v. *Simons, Barratt* v. *Woolf Seddon* and *Harris* v. *Sholfield Roberts* (*The Times*, 21 July 2000), an appellate court of seven Law Lords held that in the changed world of today immunity should no longer be available. Their Lordships were unanimous in holding that their decision should apply in civil actions, but only by a majority in extending that decision to criminal proceedings. Lord Hoffman stated:

'Public policy is not immutable and there had been great changes in the law of negligence, the functioning of the legal profession, the administration of justice and public perceptions. It was once again time to re-examine the whole matter...
...in altering the law on the immunity of advocates the House would not be intervening on matters which should be left to Parliament. The judges had created the immunity and the judges should say that the grounds for maintaining it no longer existed.'

The question whether or not an arbitrator is empowered to act inquisitorially under previous legislation has been much debated: one of the arguments in favour of that view being the wording of section 12 of the 1950 Act, under which 'the parties ... shall submit ... to be examined by the arbitrator'. Eminent authorities have however argued that adversarial procedure is deeply enshrined in English law and that this was not intended to vary that procedure. The 1996 Act expressly empowers the tribunal to act inquisitorially under section 34(2)(g), which provides, subject to the right of the parties to agree any matter, that the tribunal may decide 'whether and to what extent the tribunal should itself take the initiative in ascertaining the facts and the law'. Under section 37, again subject to the right of the parties to agree otherwise, the tribunal is empowered 'to appoint experts or legal advisers to report to it and the parties or to appoint assessors to assist it on technical matters'.

The amended appeal procedure introduced under the 1979 Act has been retained under the 1996 Act without substantial alteration. The opportunity has however been taken to codify what had become known and respected as the *Nema* rules, in section 69 of the 1996 Act; those rules providing a clear definition of the circumstances in which the court is empowered to grant leave to appeal, on a question of law arising out of an arbitrator's award.

# 1.5 *Advantages of arbitration*

Many of the advantages most frequently claimed for arbitration as an alternative to litigation are especially relevant to disputes arising from construction contracts.

## 1.5.1 Freedom to choose the arbitrator

The parties to an arbitration agreement are free to choose a suitable person to be their arbitrator. Frequently disputes arising from construction contracts involve such questions as whether or not the ground conditions encountered could reasonably have been foreseen by an experienced contractor, having regard to the subsoil information available to the contractor at the time of making the contract; whether or not the issue of drawings or instructions on certain dates caused delay to the works; whether or not variations ought to be valued at contract rates or in some other way having regard to the provisions of the contract. A proper understanding of those and many other issues likely to arise can usually be gained only by long experience in the construction industry – and preferably experience both as a contractor and as engineer under the contract. Hence it is often desirable that the arbitrator should be an experienced engineer (or, where appropriate, architect or quantity surveyor); and this objective is often achieved, if not by agreement of the parties as to their choice of the arbitrator, by naming as the appointing authority the president of the appropriate professional body. Where the parties are able to agree upon a suitable person to be their arbitrator, so much the better: but notices to concur in an appointment sometimes give rise to a suspicion – usually quite unjustified – that the author of the notice seeks to gain some unfair advantage from the proposed appointment.

While it is recognised that technical expertise is available in the courts through the appointment of experts, there is a very real danger that a non-technical judge may be influenced more by the powers of presentation and of persuasion of the expert before him than by the technical merit of his evidence.

## 1.5.2 Flexibility

Disputes arising from construction contracts may involve sums of money varying from a few thousand pounds to tens or hundreds of millions. They may involve questions of law or of fact or both, and the questions of law may arise either from the construction of standard forms of contract or from 'one-off' terms in a contract; while the questions of fact may be simply what happened during construction, or what might have been foreseen by an experienced professional. The credibility of witnesses of fact may or may not be in question. All of these matters affect the choice of an appropriate procedure and the form and level of representation, if any, of the parties.

In arbitration the parties are free to determine these matters by agreement; and while neither party can dictate to the other as to its choice of representation, a party may bring to the notice of the arbitrator a contention that costs are being incurred unnecessarily by its opponent, and may request that this be taken into account in the arbitrator's award of costs.

### 1.5.3  Economy

Critics of arbitration often argue that total costs are likely to exceed those of litigation, because in the latter the judge and court facilities are provided at public expense, while in arbitration both the arbitrator and facilities for the hearing have to be paid for by the parties. While true, this is not usually a major factor in the total costs of the proceedings, the arbitrator's fees often being much less than those of counsel appearing before him. Second, where technical issues are involved, experts may be needed to explain such issues to a technically lay judge, but not to an arbitrator having appropriate technical knowledge. And third, proceedings before a judge are likely to be more protracted, and hence more costly, than before an arbitrator having the knowledge needed to recognise the technical issues.

Economy is not however achieved automatically by the use of arbitration in preference to litigation, but it may be achieved where the parties act sensibly in choosing their arbitrator, the form of the proceedings, and their representation. Where a party wishing to act sensibly is opposed by one whose aim is otherwise – perhaps to prolong the reference and hence to defer the day of judgment – such behaviour should be brought to the notice of the arbitrator, who may exercise the powers available to him under sections 41 and 42(2)(a) of the 1996 Act for dealing with such behaviour and may take it into account in his award of costs.

## 1.5.4 Expedition

It is especially important in construction contract disputes, which often involve voluminous documentary evidence and a need to rely additionally upon oral evidence, that delay in their resolution should be avoided. Any such delay may result in documents being lost, dispersed or destroyed: while oral evidence may become less reliable because of fading memories, and dispersal or death of witnesses. Real evidence of buildings, structures, or other works may become covered or altered by later developments, adding to the difficulty of identifying and assessing alleged defects. Properly used, arbitration can provide a means whereby disputes arising from construction contracts can be resolved more readily than in litigation.

## 1.5.5 Privacy

Arbitration proceedings, unlike those in the courts, are not open to the press or to the public: only those persons involved in the proceedings are entitled to attend the hearings and meetings that are usually needed. Others not directly involved may be invited to attend by agreement of the parties; and frequently such invitations are extended to, for example, pupils of the arbitrator who wish to gain experience. Attendance is however permitted on the condition that such pupils or others will respect the confidentiality of the proceedings.

In many cases the parties to arbitration proceedings have no wish to publicise details of their dispute. Where, as is sometimes the case, a previously harmonious relationship – perhaps between a main and a subcontractor – is interrupted by a dispute, amicable relations are usually more readily restored where publicity has been avoided.

## 1.5.6 Finality

English law of arbitration has traditionally laid emphasis on compliance with the law, while in some other jurisdictions greater freedom is given to the arbitrator to dispense his own concept of justice. As a result it had, prior to the enactment of the Arbitration Act 1979, become almost standard practice in any major arbitration under English law for the losing party to pursue questions of law – sometimes spurious – through the hierarchy of the courts, thereby delaying the date of settlement; sometimes by years. The 1979 Act and subsequent case law did much to correct that abuse; and in particular the decision of the House of Lords in *BTP Tioxide* v. *Pioneer Shipping and Armada Marine (The Nema)* [1981] 2 Lloyds Rep 239 imposed firm restrictions on the court in granting leave to appeal. The 1996 Act has confirmed the limitation imposed by the *Nema* rules by codifying them within section 69 of the Act: and it has laid emphasis on the concept of fairness.

### 1.5.7 Convenience

In arbitration proceedings the parties are, subject to the agreement of the arbitrator, free to choose the dates, times and venues of hearings and of meetings. They are entitled to expect that the arbitrator will, where the parties are in agreement and where he can, comply with their wishes as to such arrangements. A measure of control by the arbitrator is of course necessary where the parties are not in agreement; and in exercising that control the arbitrator has a duty under section 33(1)(b) of the 1996 Act to avoid unnecessary delay and expense. However in most cases the parties' convenience is paramount: they may expect that dates and times will be chosen to suit their own convenience, and that dates agreed or determined by the arbitrator will not be subject to late, or indeed any, alteration.

This position may be contrasted with that in litigation, in which dates and times of hearings are determined by the court upon the application of a party, and may be subject to late cancellation or deferment. This is because the courts are aware that a large proportion, or perhaps even a majority, of civil actions are settled by negotiation, often immediately before commencement of the court hearing, and allowance is made for such settlements by double- or treble-booking court time.

## 1.6 Disadvantages of arbitration

### 1.6.1 Costs of the arbitrator and of court facilities

In contrast to litigation, where both the judge and the court facilities are provided at public expense, the parties to an arbitration, or one of them, will ultimately have to bear the costs of the arbitrator and of the courtroom facilities. However in many cases such costs are small in comparison with other costs incurred in litigation.

## 1.6.2 Unavailability of legal aid

Where arbitration is used as the means of resolving minor disputes, and in particular those in consumer industries, the unavailability of legal aid may be an important consideration. The Consumer Arbitration Agreements Act 1988 (repealed by the 1996 Act) sought to protect consumers from effects adverse to their interests where suppliers incorporate arbitration agreements in their terms of trade. In its place the Unfair Terms in Consumer Contracts Regulations 1999, which came into force on 1 October 1999, seek to protect the right of a consumer to have small claims determined in the court, notwithstanding the existence of an arbitration clause in the contract. Section 91 of the 1996 Act defines a term which constitutes an arbitration agreement as being unfair so far as it relates to a claim not exceeding an amount specified by order: currently £5,000.

Whether or not the 1999 Regulations are applicable to building contracts is however a matter of some doubt, depending upon whether or not the building owner is deemed to be a consumer. Where the building owner has sought a number of quotations for the work it is likely that he will be deemed not to be a consumer and therefore not to be protected under the 1999 Regulations.

Many minor disputes arising from construction contracts (for example those relating to building, double glazing, plastics, and heating and ventilation) are however covered by arbitration schemes administered by the Chartered Institute of Arbitrators, whose aim it is to initiate and to administer procedures under which costs are commensurate with the small sums in dispute.

## 1.6.3 Joinder difficulties

Where more than two parties are involved in related disputes – for example employer, main contractor and subcontractor – there is no statutory power whereby all parties may, as in litigation, be joined in a consolidated action or be dealt with in concurrent hearings. The question whether or not such power should be incorporated in the 1996 Act was canvassed during the consultative phases, and the replies received indicate roughly equal support for positive and negative answers to the question. Those who supported the introduction of a statutory power generally coupled that support with restrictions of one form or another. In the event the 1996 Act has included, under section 35, provision both for consolidation of arbitral proceedings with other arbitral proceedings, and for concurrent hearings, but only by agreement of the parties. The parties are however free to confer on the tribunal power to order consolidation or concurrent hearings.

Certain of the standard forms of contract – notably the CECA Form of Subcontract – provide in defined circumstances for a degree of consolidation: the form of words used in the CECA form being:

'The Contractor ... may ... require that any such dispute under this Subcontract be dealt with jointly with the dispute under the Main Contract and in a like manner.'

Provisions of this nature are however themselves likely to give rise to further difficulties and disputes. The CECA provision is enforceable by the main contractor only with the agreement of the employer: and it is far from certain, or even likely, that the employer would wish an arbitration with a main contractor to be complicated and prolonged by the inclusion of subcontract disputes. In many cases the primary objective of parties – usually main contractors – who seek consolidation or concurrent hearings is to ensure that they do not suffer from inconsistent awards. That objective should generally be achieved where the same arbitrator is appointed for both the main and the subcontract disputes, although even with that arrangement the desired outcome is by no means certain. This is because success in an arbitration depends not only upon the facts of the case, but also upon the evidence to prove those facts, and the manner of its presentation. It is far from certain that a main contractor found to be liable for a subcontractor's claims would be able to succeed in corresponding claims against the employer, even when the same arbitrator is appointed to hear both disputes and the claim is valid under the main contract (see Chapter 5 section 5.3.3).

Lord Saville, in his Foreword to this book, draws attention to some of the disadvantages of consolidation of court proceedings: in particular to the delay that a small subcontractor might have to suffer awaiting some 'blockbuster piece of litigation' years after the event. That situation has since been alleviated to some extent by the provisions of the Housing Grants, Construction and Regeneration Act 1996, under which the small subcontractor envisaged by Lord Saville should at least be able to obtain an enforceable interim decision of an adjudicator, enabling him to stay in business while awaiting a delayed arbitration. Alternatively the decision of the Court of Appeal in *Redland Aggregates* v. *Shepherd Hill Civil Engineering* (see Chapter 2, section 2.8.6) should eliminate the risk of undue delay in proceeding with a subcontract arbitration.

### 1.6.4 Incompetent arbitrators

While judges are appointed only after they have gained extensive knowledge and experience, usually at the Bar, arbitrators having inadequate qualifications and experience may be, and sometimes are, appointed either by an appointing authority or by the parties, in ignorance of the requirements of the appointment.

Many appointing authorities now maintain panels of qualified arbitrators; some requiring candidates for listing on those panels to qualify with the Chartered Institute of Arbitrators before sitting the professional body's own examination. Parties seeking to make appointments by agreement are well advised to propose only persons whose names appear in the appropriate lists and can be seen to have the necessary knowledge and experience: although even those precautions are not always sufficient.

In *Pratt* v. *Swanmore Builders and Baker* (1980) 15 BLR 37, an arbitrator appointed by the Chartered Institute of Arbitrators who had 'shown himself to be quite incompetent to conduct the arbitration' and had 'allowed the arbitration to be reduced to such a state that there was no prospect of justice being done' was removed by order of the High Court under section 23 of the 1950 Act. More recently, in *Fairclough Building* v. *Vale of Belvoir Superstore* (1990) 56 BLR 74, an arbitrator appointed by the RIBA who made three errors of law in his award in which he allowed only some £43,000 against an uncontested claim of over £400,000 had the award, with the Official Referee's judgment, remitted to him for reconsideration under section 22 of the 1950 Act. Fortunately such occurrences are rare, especially where – as for example in the case of the ICE – the appointing authority administers a rigorous examination system for aspiring arbitrators.

Lay parties, and regrettably even some appointing authorities, sometimes assume that the qualification ACIArb (Associate of the Chartered Institute of Arbitrators) implies competence to conduct arbitrations. In fact Associateship is gained by successful completion of a week-end training course, and is merely the starting point in a study of arbitration procedure and practice, normally extending over several years and leading in due course to the senior grade of Fellow. Prior to the granting to the Chartered Institute of Arbitrators of the 1999 Royal Charter and its coming into force on 22 June 1999 Fellows of the Institute who had qualified to be included in its panels of arbitrators were not distinguishable by their title from other Fellows who had not so qualified. This anomaly has now been mitigated by provisions within the 1999 Royal Charter under which

(a) Fellows who have qualified as being competent arbitrators have become entitled to use the designation Chartered Arbitrator; and (b) in future members will not be graded as Fellows until they have so qualified. Those who have already been graded as Fellows without having fully qualified as arbitrators will remain entitled to use that designation, but 'unqualified' Fellows in this category will over a period of years be reduced in numbers, as they either qualify as arbitrators or cease to be members of the Chartered Institute.

The new designations should be helpful to parties seeking to appoint an arbitrator by agreement and to appointing authorities, by providing a clearer identification of those members of the Chartered Institute who have fully qualified with that body as arbitrators. However until such time as all of the existing 'unqualified' Fellows have either qualified as arbitrators or have ceased to be members, those seeking to appoint an arbitrator should ensure that their candidate is a Chartered Arbitrator.

# CHAPTER TWO
# ARBITRATION AGREEMENTS

## 2.1  Synopsis

An arbitration agreement, which may be entered into before or after a dispute arises, must be in writing, or otherwise recorded, in order that the provisions of the 1996 Act shall apply. It remains effective even if the contract within which it is contained is or becomes invalid or does not come into existence and it may include agreements on procedural matters. Most of the standard forms of construction contract incorporate arbitration agreements.

## 2.2  Definition

Section 6(1) of the 1996 Act defines an arbitration agreement as meaning

'an agreement to submit to arbitration present or future disputes (whether they are contractual or not)';

and section 6(2) provides that

'the reference in an agreement to a written form of arbitration clause or to a document containing an arbitration clause constitutes an arbitration agreement if the reference is such as to make that clause part of the agreement'.

Under section 5 of the 1996 Act the provisions of Part I of the Act (for practical purposes the whole Act) apply only where the arbitration agreement is in writing. An agreement in writing is deemed to exist (a) whether or not the written agreement is signed by the parties; (b) where the agreement is made by exchange of written communications; and (c) where the agreement is evidenced in writing. Additionally there is, under section 5(3) of the 1996 Act, an agreement in writing where the parties agree otherwise than in writing to written terms or to a document containing an arbitration clause; and an agreement is evidenced in writing where an oral agreement is recorded by one of the parties, or by a third party, with the authority of the parties to the agreement. Again, an exchange of written submissions in arbitral or legal proceedings in which the existence of an oral agreement is alleged by one party and is not denied by the other party in his response constitutes an agreement in writing. And finally, references to an agreement being 'written' or 'in writing' include its being recorded by any means.

## 2.3  *Separability*

An arbitration agreement is separable from the agreement to which it relates, such that it does not become invalid, non-existent or ineffective where the agreement to which it relates is invalid or ineffective or does not come into existence (section 7 of the 1996 Act).

## 2.4  *Agreements to refer*

Arbitration agreements made before a dispute arises, including those incorporated in the many forms of construction contract such as the ICE, FIDIC and JCT forms, and the CECA form of sub-contract, are often termed 'agreements to refer' because they provide for reference to arbitration of any dispute that may later arise from the contract.

## 2.5  **Ad hoc** *agreements*

Where a dispute arises from a contract in which there is no arbitration agreement within the meaning of section 6 of the 1996 Act it is open to the parties to enter into an *ad hoc* arbitration agreement in respect of the dispute (see for example SD/1 in Appendix A). In practice however it is sometimes difficult for one party to persuade the other to enter into such an agreement once a dispute has arisen.

## 2.6  *Stay of court proceedings*

Where a party to an arbitration agreement chooses to ignore that agreement and commences proceedings in court, it is open to the other party to apply to the court for a stay of the proceedings, under section 9(4) of the 1996 Act. Such application (see for example SD/2 in Appendix A) must however comply with certain requirements, namely:

(1)  the applicant must acknowledge the legal proceedings against him and must notify the other party of his application to the court

(2)  the applicant must not take any steps in the legal proceedings in answer to the substantive claim.

The existence in the contract of a requirement that disputes must be referred to other dispute resolution procedures (such as the conciliation procedure incorporated in the sixth and seventh editions of the ICE Conditions of Contract) before resorting to arbitration does not inhibit an application under section 9(4) of the 1996 Act.

The court has a duty to grant such a stay unless it is satisfied that the arbitration agreement is null and void. However in the event that the court refuses to grant a stay any provision in the contract that an award is a condition precedent to the bringing of legal proceedings (that is, a *Scott* v. *Avery* clause) is of no effect.

Provision is made in section 86(2) of the 1996 Act for the court to have discretion as to whether or not to grant such a stay in the case of domestic arbitrations, where there are 'sufficient grounds for not requiring the parties to abide by the arbitration agreement'. That section has not however been brought into force, and it appears probable that in due course it will be repealed. Currently the court has the duty under section 9 referred to above: namely to grant a stay in all cases except where the arbitration agreement is null and void.

## 2.7  Procedural matters

Having agreed to refer to arbitration any disputes that may arise from a contract between them the parties are free, under the 1996 Act, to agree as to the procedures to be adopted should a dispute arise. The Act defines, in Schedule 1, the mandatory provisions. Other provisions, termed the 'non-mandatory provisions', include certain fall-back provisions which apply in the absence of agreement between the parties, as follows.

### 2.7.1  Constitution of the tribunal

Under section 15, the tribunal shall consist of a sole arbitrator unless the parties agree otherwise. If the parties are so perverse as to agree upon two arbitrators, or any other even number, then the Act provides for the appointment of an additional arbitrator as chairman. Detailed provisions are included, under sections 16, 17, 18, 20, 21, and 22, for the various failures of appointment procedures and other difficulties that may arise, especially where the tribunal consists of more than one person.

Fortunately the usual practice in the case of domestic arbitrations arising from construction contracts to which English law applies as the arbitration procedural law, is to appoint a sole arbitrator. No action is necessary in order to implement such an intention, because it is implemented automatically in the absence of an agreement to the contrary. In such cases much of the complication envisaged in the above sections is avoided. The parties are free, when a dispute arises and is referred to arbitration, to agree upon the name of the sole arbitrator or, failing agreement within 28 days of the service of a request by one party to agree, to apply to the appointing authority where one is named in the agreement. If the parties fail to name an appointing authority in their agreement, then the court is empowered, under section 18 of the 1996 Act, to give directions as to the making of any necessary appointments. In so directing the court shall, under section 19, have regard to any agreement of the parties as to the qualifications required of the arbitrator.

Arbitration agreements included in most, if not all, of the standard forms of construction contract other than those relating to international contracts, are satisfactory in that they provide for the appointment of a sole arbitrator, and they name the authority (the president of the relevant professional institution) empowered to appoint the arbitrator in the absence of agreement. Hence no further agreement is needed where such forms apply.

### 2.7.2  Filling vacancies

Subject to the parties' agreement otherwise, detailed provisions are made in section 27 of the 1996 Act for filling any vacancy that may arise from the arbitrator's ceasing to hold office. In general, the fallback provisions of the section reflect those provisions made for the original appointment, and require no special modification. The circumstances in which a need may arise to fill a vacancy: namely where the arbitrator's authority is revoked by the parties, or where the arbitrator resigns or dies or ceases to have jurisdiction, are dealt with in Chapter 4.

### 2.7.3 Procedural and evidential matters

Section 34 of the 1996 Act empowers the tribunal 'to decide all evidential and procedural matters, subject to the right of the parties to agree any matter'. Most of the matters referred to in the section are best determined after a dispute has arisen, and they are dealt with in Chapter 5. There are however two matters, applicable to international contracts, for which decisions of the parties are best included in the arbitration agreement. Those matters are the choice of location of any part of the arbitration proceedings, referred to in section 3 of the 1996 Act as 'the seat of the arbitration'; and the language or languages to be used in the proceedings; referred to in section 34(2)(b) of the 1996 Act. The parties are of course free to agree such matters and should incorporate their agreement in the arbitration agreement.

### 2.7.4 Exclusion of right to appeal on a question of law

Sections 45 and 69 of the 1996 Act, which give a limited right of appeal to the court on questions of law arising during the proceedings (section 45) or from the award (section 69), may be excluded by agreement of the parties: such an agreement being termed an 'exclusion agreement' (see SD/3 in Appendix A). Provision is made, in section 87 of the 1996 Act, for an exclusion agreement to be void, in the case of a domestic arbitration agreement, unless entered into after the dispute has arisen. That section, together with sections 85 and 86, has not however been brought into force with the remaining sections of the Act; and hence there is at present no distinction as between domestic and international contracts in respect of the right to exclude the jurisdiction of the courts to determine questions of law arising either during the course of a reference or from the award. An agreement to dispense with reasons for the arbitrator's award is deemed, under section 45 of the 1996 Act, to constitute an exclusion agreement.

## 2.8 *Arbitration agreements in standard forms of contract*

Arbitration and other forms of dispute resolution procedures defined in standard forms of construction contract have in the closing years of the twentieth century been affected by the 1996 Arbitration Act and by the Housing Grants, Construction and Regeneration Act 1996 (the Construction Act). In addition the Civil Procedure Rules 1998 (CPR), which implement the Woolf Reforms, became effective on 26 April 1999 and have introduced new concepts in such areas as the management by the tribunal of litigation and of alternative dispute resolution (ADR) procedures. Although the CPR are primarily directed at litigation, they are also expressly relevant to arbitration.

Those responsible for drafting standard forms of construction contract have responded to the statutory changes, generally by amending the rules relating to dispute resolution procedures within such forms. Amendments published initially as separate documents have later been incorporated in revised editions of the form of contract to which they relate. However even where revision of the standard forms has not kept pace with legislative changes the latter will prevail. The 1996 Arbitration Act, for example, is applicable to all arbitrations commenced on or after 31 January 1997, irrespective of the date of the arbitration agreement. Similarly the Construction Act, which requires that construction contracts, as defined therein, incorporate a right of either party to refer any dispute that may arise to adjudication (again as defined), and gives parties to construction contracts certain rights as to stage payments, includes a provision that where a construction contract does not incorporate such rights, the adjudication and payment provisions of the Scheme for Construction Contracts (England and Wales) Regulations 1998, SI 1998 No 649 (the Scheme) which came into force on 1 April 1998, shall apply.

Although adjudication pursuant to the contract or to the Scheme, as the case may be, is not directly related to arbitration, it forms an important part of dispute resolution procedure in construction contracts. As such it is dealt with in Chapter 11, section 11.6, in the context of dispute management.

Both the ICE and the JCT families of contract forms include arbitration agreements which vest in the arbitrator an additional power, quite distinct from his statutory powers, '...to open up review and revise any decision opinion instruction direction certificate or valuation of the Engineer or an adjudicator'. Slightly different wording is used in the JCT forms, but the power (which in the JCT forms refers to decisions of the architect) is essentially the same. The construction of the words used has formed the subject of many important judgments during the period from 1984 to 1998: see Chapter 4, section 4.7.6.

## 2.8.1 The ICE Conditions of Contract Measurement Version, seventh edition

Clause 66 of the current (seventh) edition of the ICE Conditions of Contract, published in September 1999, has been substantially amended in order to comply with the requirements of the Construction Act, and to promote the avoidance and settlement of disputes by arbitration or by other means. In addition the use of the ICE Arbitration Procedure (1997), mandatory under the fifth and sixth editions, has been replaced by an option to adopt either that procedure or the Construction Industry Model Arbitration Rules (see subsections 2.8.12 and 2.8.13). That option is impliedly to be exercised by the Employer and is implemented by way of an insertion in paragraph 22 of the Form of Tender (Appendix).

Clause 66 includes complex and detailed provisions for adjudication pursuant to the Construction Act, for determination of disputes by the engineer, for conciliation, and finally for arbitration. Hence it necessarily includes an arbitration agreement within the meaning of section 6 of the 1996 Act, which agreement provides for the appointment of an arbitrator, failing agreement by the parties, by the President or a Vice-President of the Institution of Civil Engineers. The arbitration agreement is heavily circumscribed by provisions intended to promote settlement of the dispute before arbitration is invoked and to provide the arbitrator with additional powers during his conduct of the reference.

Subclause 2 of clause 66 provides for reference by the employer or the contractor of any matter of dissatisfaction to the engineer for his decision, which shall be notified to both parties within one month of the reference.

Subclause 3 of clause 66 provides that no matter shall constitute nor be said to give rise to a dispute unless and until in respect of that matter the engineer has given a decision under subclause 2 which is unacceptable to the employer or the contractor, or the time for giving such decision has expired, and the employer or the contractor has served on the other and on the engineer a notice in writing (called the Notice of Dispute).

Following upon the service of a notice under subclause 3 both parties are required under subclause 4 to give effect to the engineer's decisions, notwithstanding that such decisions may be disputed; and to continue to perform their obligations. Where the dispute arises from the decision of an adjudicator appointed pursuant to subclause 6 (see below) it is similarly effective.

Subclause 5 provides, where both parties so agree, for an attempt to resolve the dispute by conciliation under the ICE Conciliation

Procedure 1999; and where the conciliator makes a recommendation, it is deemed to have finally determined the dispute by agreement unless challenged by either party within one month of receipt of the recommendation.

Subclause 6 provides for adjudication in accordance with the ICE Adjudication Procedure 1997, which procedure is designed to satisfy the requirements of the Construction Act. It provides for the appointment of an adjudicator within seven days of the service of a Notice of Adjudication by one party upon the other, and for the adjudicator's decision to be given within 28 days of the date of referral to him. This period may however be extended by up to 14 days by the adjudicator, subject to the consent of the referring party. The procedure has the merit of providing a rapid interim resolution of disputes that may arise, typically from an alleged failure of one of the parties to a construction contract to make payment in accordance with the terms of the contract or where applicable with the Construction Act, and has proved to be effective where a party defaults in making payment (see Chapter 11, section 11.6).

Under subclause 66(9) disputes that remain in existence after the foregoing procedures of conciliation and/or adjudication have either been rejected or have proved to be unsuccessful may be referred to arbitration by the service by one party on the other of a Notice to Refer, which notice must be given within three months of a decision by an adjudicator where that decision forms the subject of the dispute, failing which the adjudicator's decision becomes final and binding.

Subclause 66(10) of the ICE conditions provides for appointment of the arbitrator by agreement of the parties, with the proviso that in the event of failure to reach such an agreement within one month of the issue by one party to the other of a Notice to Concur either party may apply to the President of the ICE for the appointment of an arbitrator. The subclause also includes provision for a Vice-President of the ICE to act where the President is unable to exercise his function, and for a replacement to be appointed by a similar procedure where the first arbitrator declines the appointment, is removed by the court, or dies.

The ICE publishes a List of Arbitrators giving the names, details and brief *curricula vitae* of the persons listed: namely those persons aged under 72 who have qualified with the Institution as arbitrators. In order to qualify candidates must have achieved the standard required by the Chartered Institute of Arbitrators for qualification as a Chartered Arbitrator and have satisfied the ICE as to their knowledge of the law relating to construction contracts and the ICE forms of contract in particular. For that purpose the candidate is required to pass a written examination (termed the Endorsement Examination) and to satisfy the examiners at an interview.

The ICE List of Arbitrators, which is updated at frequent intervals, provides a valuable summary of the knowledge and experience, both in civil engineering and in arbitration, of the persons listed, and as such is a useful reference for parties seeking to appoint an arbitrator by agreement.

Subclause 66(11) confirms that the arbitration agreement is pursuant to the Arbitration Act 1996 and that the procedural rules shall be either those of the ICE Arbitration Procedure (1997) or the Construction Industry Model Arbitration Procedure (CIMAR) (see subsections 2.8.12 and 2.8.13).

Provision is made in section 67 of the ICE form for application to arbitrations in Scotland and in Northern Ireland. In the former case, the contract is to be construed in accordance with Scots law: 'arbitrator' becomes 'arbiter', and the Arbitration Act 1996 is to be replaced by a reference to 'the law of Scotland and/or section 66 and schedule 7 of the Law Reform (Miscellaneous Provisions) (Scotland) Act 1990 as may be appropriate'. The relevant procedural rules are the ICE Arbitration Procedure (Scotland) (1983) or later edition of those rules.

In the case of works in Northern Ireland, the contract is to be construed as a Northern Irish contract and interpreted in accordance with the law of Northern Ireland. The Arbitration Act 1996 does of course apply in Northern Ireland.

## 2.8.2   The ICE Conditions of Contract for Minor Works, third edition

First published in 1988, revised as a second edition in 1995 and reprinted with revisions in 1998, the ICE Minor Works Conditions was republished as a third edition in 2000. Revisions since the original second edition take account principally of the Arbitration Act 1996, the Construction Act, and the replacement of the former Federation of Civil Engineering Contractors (FCEC) by the Civil Engineering Contractors' Association (CECA). In addition the original disputes procedure contained within clause 11.1 of the conditions has been replaced in its entirety by an Addendum A entitled Avoidance and Settlement of Disputes, the content of which is identical to that of the corresponding procedure incorporated in clause 66 of the ICE Conditions of Contract (see subsection 2.8.1).

One of the consequences of this change is that a distinguishing feature of the original minor works form in comparison with the ICE conditions, namely the omission of reference of a dispute to the engineer before invoking arbitration or another dispute resolution procedure, no longer applies. Both forms now require such reference as a preliminary to invoking conciliation, and/or adjudication, and/or arbitration.

Adoption of the arbitration procedure incorporated in clause 66 of the ICE conditions has also resulted in the correction of a drafting error in the original procedure. Clause 11 originally specified that the arbitration be conducted in accordance with the Short Procedure of the ICE Arbitration Procedure (1983). That procedure includes a provision in rule 21.1 that the arbitrator's costs be shared equally between the two parties: a provision which was rendered ineffective under section 18 of the Arbitration Act 1950 and remains so under section 60 of the 1996 Act.

Where a contract under the Minor Works Conditions has its seat in a jurisdiction outside England, Wales, Scotland or Northern Ireland (for example in the Channel Islands or the Isle of Man), or where the Construction Act does not apply, the Guidance Notes for the Conditions provide a formula for simplifying the dispute resolution procedure outlined above. This provides for replacement of clause 11.1 and addendum A by the disputes procedure included in the Second Edition of the Minor Works Conditions, with slight updating. In addition the drafting error referred to above has been corrected by requiring the parties, if they wish the Short Procedure of the ICE arbitration procedure which includes rule 2.1 to apply, to enter into an agreement to that effect after service of the notice of dispute.

### 2.8.3 The ICE Design and Construct Conditions of Contract, second edition

First published in October 1992 the D&C conditions were supplemented by a Guidance Note issued in March 1995, and were amended in February and March 1998 in order to take account of various statutory changes. The conditions recognise the increasing popularity of the 'package deal' type of contract, in which the contractor is responsible for both design and construction. Under such a contract the contractor often engages the services of a consulting engineer, thereby providing the level of design skill expected in traditional forms of construction contract, where the engineer is usually employed under a separate contract. An important difference however is that under the D&C form the engineer's client is

23

the contractor instead of the employer: and this relationship avoids the isolation or even antagonism that sometimes exists between engineer and contractor in conventional forms of contract. Furthermore in the D&C form, the employer is not concerned with many of the sources of dispute that often arise under traditional forms of contract: for example delays caused by the late issue of design information, design errors, errors in setting out by the engineer and in the measurement of quantities, and the cost of variations and remedial works where they do not originate from variations ordered by the employer. All such delays and their resulting costs become the responsibility of the contractor, and therefore are an internal matter within the contractor's organisation. Hence the scope for claims by the contractor against the employer is greatly reduced, although such claims may of course result in disputes between the engineer and the contractor under the separate contract covering that relationship.

Although the inclusion of design responsibility in what would otherwise be a conventional form of contract constitutes a major change from that conventional form, the format of the D&C is remarkably similar to that of the ICE conditions of contract sixth edition, from which it has been developed. Its authors have wisely retained the clause numbering system that has been common to most if not all of the successive editions of the ICE conditions almost in its entirety. Numbers of those clauses of the sixth edition which became redundant in the D&C form were marked as being 'not used'.

The detailed and complex provisions of clause 66 of the ICE seventh edition relating to dispute resolution in its many forms (see section 2.8.1) are reproduced in their entirety in clause 66 of the D&C form.

### 2.8.4 The NEC Engineering and Construction Contract, second edition

First published in 1993, the New Engineering Contract in its second edition is entitled the NEC Engineering and Construction Contract but is generally known by its original title or its acronym NEC. It appeared in 1995 and was revised in 1998 to take account of the Construction Act. It comprises a number of 'core clauses' covering matters common to all of the 'main option clauses' which follow. The first four options are either a priced or a target contract combined with either an activity schedule or a bill of quantities. Two further options are a cost reimbursable contract or a management contract.

The language of the form avoids the mandatory terms used in most forms of contract, and instead uses the present tense to express what appear to be expectations rather than mandatory requirements: for example 'the project manager issues his certificates' and 'payments are made by the employer to the contractor'.

The dispute resolution procedure forms the final section of the core clauses, and was updated in April 1998 by an Addendum to take account of the Construction Act in cases where the contract is subject to that Act. It comprises an initial 'notice of dissatisfaction' followed within two weeks of the notice by attendance at a meeting between the contractor and the project manager in an attempt to resolve the matter. Provision is made for a further meeting, with the same objective, where either party is dissatisfied with 'any other matter'. However either party may at any time give notice of his intention to refer a dispute to adjudication, the procedure and timing of which follows the Scheme introduced by the Construction Act. Thereafter a party may apply, within four weeks of notification of the adjudicator's decision or his failure to give such decision within the time prescribed in the Construction Act, to the 'tribunal'. The adjudicator is appointed by the employer and is named in the contract.

The NEC form provides for the employer to specify 'arbitration' as being the 'tribunal', or to leave that term undefined; implying the tribunal to be the court. Both the adjudicator and the tribunal are empowered to 'review and revise any action or inaction of the Project Manager or Supervisor'.

## 2.8.5 The FIDIC Conditions of Contract for Construction

The FIDIC Conditions of Contract (International) for Works of Civil Engineering Construction was first published by the Federation Internationale des Ingénieurs-Conseils in 1957. The second, third and fourth editions appeared in 1969, 1977 and 1987 respectively, and various reprints, some with amendments, have appeared from time to time. The current edition, now entitled the Conditions of Contract for Construction: First Edition was published in 1999 and is, as its name implies, a completely new document, expressly intended for 'building and engineering works designed by the employer'. Other editions of the conditions in the series are published covering Plant and Design-Build, Engineering Procurement and Construction (EPC)/Turnkey Projects, and a Short Form of Contract. Unlike its predecessors the new FIDIC conditions do not follow the ICE forms in content or in format.

The dispute resolution procedure initially comprises reference to a Dispute Adjudication Board consisting of either one or three members, appointed jointly by the parties. In the absence of definition in the contract the DAB comprises three members. A decision, with reasons, in respect of disputes referred to the DAB must be given within 84 days of referral, subject to variation by agreement with the parties, and is binding upon the parties unless and until revised by 'amicable settlement' or by an arbitral award. Notice of dissatisfaction with the DAB decision must be given within 28 days of the decision or, where the decision has not been given within the prescribed period, within 28 days of the expiry of that period. In the event of failure to give such notice the decision becomes final and binding upon both parties.

Where the decision has not become final and binding or has not been given, the dispute is referred to arbitration under the rules of the International Chamber of Commerce. The tribunal, of three arbitrators unless otherwise agreed, has full power 'to open up, review and revise any certificate, determination, instruction, opinion or valuation of the Engineer and any decision of the DAB relative to the dispute'. The parties are not limited in the proceedings before the tribunal to the evidence or arguments previously put before the DAB to obtain its decision, or to the reasons for dissatisfaction given in its notice of dissatisfaction.

Under the ICC Rules arbitrators' fees are based on a scale related to the sum in dispute and to the exercise of discretion by the court of the ICC. The fee for a sole arbitrator varies widely from a minimum of US$ 2,500 to a maximum of 17% of the disputed sum where that sum does not exceed US$ 50,000. Where the disputed sum exceeds US$ 50,000 the percentage reduces progressively from a range of 2% to 11% where the disputed sum is $50,000 to a minimum of between 0.01% and 0.05% where the disputed sum exceeds US$ 100m. In addition the arbitrator is entitled to his personal expenses. Where the tribunal consists of three arbitrators, the ICC Court may at its discretion increase the total fee to a maximum of three times the fee payable to a sole arbitrator. In addition the ICC Court charges for its administrative expenses on a scale which provides generally for a charge of approximately one half of the sole arbitrator's fee. Although the scale provides for a substantial degree of flexibility, at the discretion of the ICC Court of Arbitration, which must have regard to the time spent, the rapidity of the proceedings and the complexity of the dispute, it is likely that in many cases the total cost of an arbitration under the ICC Rules will be substantially greater than where a sole arbitrator is appointed and where arbitration procedure is governed by English law, namely the 1996 Act.

### 2.8.6 The CECA Form of Subcontract

The demise of the Federation of Civil Engineering Contractors and its replacement by the Civil Engineering Contractors Association has resulted in a corresponding change in the sponsorship of the well known and respected Form of Subcontract. Current versions of that document, published in July, October and November 1998, are intended for use in conjunction with main contracts under the ICE sixth edition, the ICE fifth edition, and the ICE design and construct contract, respectively. The disputes procedure, under clause 18 of each of the forms, is similar and takes account of the provisions of the Arbitration Act 1996 and of the Construction Act 1996.

Clause 18 provides for optional conciliation under the ICE Conciliation Procedure (1994), for adjudication under the ICE Adjudication Procedure (1997) (which procedure is intended to comply with the requirements of the Construction Act), and where neither of these procedures results in a final determination, for arbitration.

Subclause 18(10) of the CECA form provides in certain circumstances for a degree of consolidation of subcontract disputes with disputes arising under the main contract, where such disputes have been referred to conciliation, adjudication, or arbitration. The main contractor is empowered, where a main contract dispute has been referred to conciliation or adjudication, to require that the subcontract dispute be referred to the conciliator or adjudicator to whom the main contract dispute has been referred. The main contractor is also empowered, where a main contract dispute has a connection with a dispute under the subcontract, to require the subcontractor to provide information and attend meetings in connection with the main contract dispute.

Where the subcontractor wishes to refer to arbitration a dispute that has arisen under the subcontract, and the main contractor considers that the dispute raises a matter which he wishes to refer to arbitration under the main contract, the main contractor may, under clause 18(10), require that the subcontract dispute be determined jointly with the arbitration to be commenced under the main contract. That provision, which repeats similar requirements in earlier versions of the form as published by the FCEC, has given rise to several difficulties in the past.

One of those difficulties is that while provision is made in section

35 of the Arbitration Act 1996 for consolidation of proceedings or for concurrent hearings, such procedures may be adopted only with the agreement of the parties. Where, as in the common situation, a main contractor seeks to consolidate a dispute under a subcontract with another under the main contract, he can do so only with the agreement of the employer. In many such cases the employer will have no wish to complicate and probably to prolong the proceedings, at additional cost, by allowing a third party to enter into those proceedings, and is therefore likely to withhold his agreement to any form of consolidation.

Another difficulty that often arises is that of delay resulting from any attempt to consolidate arbitration proceedings. A subcontractor may not wish to have his dispute with the main contractor held in abeyance, sometimes for several years, until the main contract work has been completed and all disputes under the main contract have been through the various stages of reference to the engineer, conciliation and reference to arbitration, and it would be unfair to expect the subcontractor to do so. Fortunately that defect in the form of subcontract has now been remedied by the court.

In *Redland Aggregates* v. *Shepherd Hill Civil Engineering* 1999 BLR 252, the Court of Appeal considered the operation of clause 18(2) of the FCEC 'Blue Form', in which similar provision was made for consolidation of sub- and main contract disputes to that in clause 18(10) of the current CECA form. Sir Christopher Staughton in his leading judgment stated:

'I have no hesitation in holding that there is implied in the clause a term that, if the Contractors are unable or unwilling to bring about a tripartite arbitration within the time span contemplated by the clause, the Sub-Contractors will be free to proceed with an arbitration under Clause 18(1). That seems to me necessary to give business efficacy to the contract.'

Earlier in the judgment Sir Christopher had stated:

'But in so far as the progress of a tripartite arbitration depends on the Contractors, I consider that it ought to be set up and conducted with all deliberate speed.'

The Court of Appeal held that the subcontractor was entitled to pursue his arbitration separately, without having to await the consolidated action sought by the main contractor.

### 2.8.7 The JCT Standard Form of Building Contract (JCT98)

The Standard Form of Building Contract published by the Joint Contracts Tribunal is currently in an edition dated 1998 and is commonly known as JCT98. Since that date there have been amendments dated June 1999 and January 2000, and the form has been reprinted incorporating them. The form is published in 6 versions, covering all of the permutations of Local Authorities or Private and With Quantities, Without Quantities, or With Approximate Quantities.

The articles of agreement provide for disputes to be resolved by adjudication, and by arbitration or by legal proceedings. Clause 41A of the contract provides for adjudication under a procedure designed to satisfy the requirements of the Construction Act. The adjudicator is to be nominated by the president or a vice-president or chairman of the RIBA, the RICS, the Construction Federation or the National Specialist Contractors Council, the choice of nominating body being stated in the appendix to clause 41A.2. In the event that the specified body fails to nominate, the party initiating the adjudication may select any of the other nominating authorities to perform that function.

The appendix also includes the statement 'clause 41B applies', indicating that arbitration is the chosen tribunal for disputes not referred to adjudication, or arising from an adjudicator's decision. Where the parties opt for litigation, the statement 'clause 41B applies' is to be deleted. Clause 41B provides for the arbitrator to be appointed by president or a vice-president of the RIBA, the RICS or the CIArb, the parties' choice being indicated by deletion of two of the named authorities. Where the parties fail to make such deletions the fall-back appointing authority is the RIBA.

Where the dispute is referred to arbitration, it is to be conducted under CIMAR (see section 2.8.13 below) current at the *Base Date*, which is defined as the date entered in the Appendix.

### 2.8.8 The JCT Agreement for Minor Building Works (MW98)

First published in January 1980, the JCT Form of Agreement for Minor Building Works is intended for use where minor building works are to be carried out for an agreed lump sum. The latest revision of the form was published in 1998, since when two amendments, MW1 and MW2 have been issued in 1999 and 2000 respectively. A Guidance Note states that the form is generally suitable for building work (new, extensions, refurbishment etc.) up to the value of £100,000 at 2000 prices.

A simple arbitration agreement is incorporated as clause 8.2 and article 7A of the form, providing for written notice by either party requiring a dispute to be referred to arbitration: and providing that, failing agreement of the parties within 14 days of the notice, the arbitrator shall be appointed by the authority named in the contract. Provision is made for that authority to be the President of the RIBA, the RICS or the CIArb. The arbitrator so appointed is empowered

> 'to rectify this Agreement so that it accurately reflects the true agreement made between the parties, to direct such measurements and/or valuations as may in his opinion be desirable in order to determine the rights of the parties and to ascertain and award any sum which ought to have been the subject of or included in any certificate, and to open up, review and revise any certificate, opinion, decision, requirement or notice and to determine all matters in dispute which shall be submitted to him in the same manner as if no such certificate, opinion, decision, requirement or notice had been given.'

As in the case of the JCT Standard Form, the arbitration is required to be conducted in accordance with the Construction Industry Model Arbitration Rules (CIMAR) current at the date of the Agreement (see section 2.8.13).

Unusually however the clause includes an agreement that either party may appeal to the High Court on a question of law arising out of an award, or may apply to the High Court for determination of any question of law arising in the course of the reference. This agreement refers to sections 1(3)(a) and 2(1)(b) of the 1979 Arbitration Act, which correspond to sections 69(2)(a) and 45(2)(a) respectively of the 1996 Act. It is in effect the opposite of an exclusion agreement, in that it paves the way for an application or an appeal, seeking to transfer from the arbitrator to the court jurisdiction to determine questions of law. It is however unlikely that an application for leave to appeal (which is required in either case) would be successful unless there were valid reasons for allowing it: which reasons would in many cases justify the court's intervention on the application of one party, even had there been no such agreement.

## 2.8.9 The JCT Intermediate Form of Building Contract (IFC98)

First published in 1984, the JCT Intermediate Form has since been updated on several occasions by way of revisions, and was republished as the 1998 edition (IFC98) in that year. Further amendments numbers 1 and 2 have since been published, in 1999 and 2000 respectively. The form is intended for use for works of simple content, in the range between those for which the JCT standard form of building contract and the JCT agreement for minor building works is issued. It makes no provision for nomination, but provides for the naming of approved subcontractors, who are to be employed as domestic subcontractors where approval is given by the architect or the contract administrator to their employment.

The dispute procedure is similar to that incorporated in the JCT minor works form (see subsection 2.8.8).

## 2.8.10   The Institution of Chemical Engineers' Form of Contract

Published as a third edition in 1995, the Institution of Chemical Engineers' form of contract for process plant (The Red Book) is expected to be updated in mid-2001 as a fourth edition, in which clause 45 will deal with disputes in general; clause 46 with a procedure for reference of disputes to an expert; and clause 47 with arbitration.

Clause 45 will provide initially for matters with which the contractor is dissatisfied to be referred, with full details, to the Project Manager, who is required to give his written decision on the matter within 28 days of the reference. If the decision is not given within 28 days, or is not implemented within 21 days of its issue, or is unacceptable, it is deemed to constitute a dispute, of which either party may give notice. The parties must then attempt to resolve the dispute by negotiation, or by mediation in accordance with procedures of the Centre for Dispute Resolution (CEDR) or some other body. Where neither of these courses leads to settlement, the dispute may be referred to an Expert appointed pursuant to clause 46 or to arbitration pursuant to clause 47.

Expert determination under clause 46 provides for the expert to be appointed, failing agreement and upon the application of either party, by the Institution of Chemical Engineers, and for the reference to be conducted in accordance with the IChemE Rules for Expert Determination. The expert is required to decide all disputes referred to him as an expert and not as an arbitrator. He may overrule any decision or instruction of the project manager, and may decide both factual issues and the interpretation of the contract. The procedure implements a principle propounded by Lord

Hoffman in *Beaufort Developments (NI) Ltd* v. *Gilbert Ash NI Ltd (1998)* (see Chapter 4, subsection 4.7.6).

Arbitration pursuant to clause 47 of the contract provides, in subclauses 47.1 and 47.2, for a sole arbitrator to be appointed, failing agreement within one month of either party serving a notice to concur and upon the application of either party, by the IChemE. It is, under subclause 47.3, to be conducted in accordance with the IChemE arbitration procedures and the arbitrator has the power to open up, review, revise or overrule any decision of the project manager other than any decision expressly stated in the Contract to be final and binding. The award of the arbitrator is final and binding.

Clause 47.4 includes a provision for reimbursement to the contractor of any loss or additional expense incurred in performing the contract where a decision of the project manager which the contractor notifies that he disputes is later overruled by the arbitrator.

The exclusion of the arbitrator's power under subclause 47.3 to overrule decisions of the project manager where such decisions are expressly stated in the contract to be final and binding may appear to be in conflict with the rules of natural justice, but is not so. The contract makes provision, in clause 17, for the Project Manager to approve, modify or reject variations proposed by the contractor, and the project manager's decision on such matters is expressly 'final, conclusive and binding and not capable of being reviewed, revised or reversed by reference to an Expert under Clause 46 ... or resolution under clause 45 (Disputes)'. Hence the limitation to the arbitrator's powers to 'open up, review, revise or overrule' decisions of the Project Manager relates to engineering decisions and not to matters relating to payment.

The Project Manager is expressly required, under clause 11, to 'exercise his discretion or make a judgement or form an opinion ... to the best of his skill and judgement as a professional engineer and shall be impartial as between the Purchaser and the Contractor'.

The provision of clause 47.4 referred to above appears to be unnecessary, since the arbitrator would even in the absence of such provision be empowered to deal with the financial consequences of his overruling a decision by the project manager.

### 2.8.11  The joint IMechE/IEE Model Forms of General Conditions of Contract

First published by the Institution of Electrical Engineers in 1903, plant forms appeared long before most of the standard forms of contract in use in the construction industry, including those of the ICE, which first appeared in 1945. In later editions the IEE was joined by the Institution of Mechanical Engineers, and in 1952 the Association of Consulting Engineers agreed to recommend the use of the form to its members, from which date the ACE was included on the title page. The current edition of the form for use in home and overseas contracts for the supply only of electrical, electronic or mechanical plant is dated 1999, and carries the acronym MF/2 (rev 1). Where the contract includes both supply and erection the relevant form is dated 1988 with amendments dated 1989, 1992 and 1995 and supplements dated 1998 and 1999, and has the acronym MF/1 (rev 3). A consolidated 2000 edition (MF/1 (rev 4)) is being drafted.

The procedure for dealing with disputes commences with a brief subclause, numbered 2.6, under which the contractor may give the engineer written notice of dispute, with reasons, of any written decision, instruction or order, within 21 days of its receipt. The engineer must, within a further 21 days, confirm, reverse or vary his decision, instruction or order, notifying both the contractor and the purchaser.

If either party disagrees with the engineer's instruction etc. as confirmed, he may give notice of referral to arbitration within a further 21 days. In the absence of such notice the engineer's decision, instruction or order becomes final and binding upon both parties.

Clause 37 of form MF/2 (rev 1) defines the arbitration procedure, under which the arbitrator is to be appointed, failing agreement, by the president of the institution named in the appendix, which includes the IMechE or the IEE as the appointing authority. The arbitrator's powers include the usual power to 'open up, review and revise' engineers' decisions, as in many other forms of construction contract, which power is also available to the court (see *Beaufort Developments (NI) Ltd* v. *Gilbert Ash NI*: Chapter 4, subsection 4.7.6). They also include the power to order provisional relief, activating the provisions of section 39 of the 1996 Act, to which the parties' agreement is given in the clause.

Another power included in subclause 37.4 of the arbitration clause provides for the joinder of a subcontractor as a party to the arbitration, where there is a related subcontract dispute. In such cases the parties also agree to the arbitrator's appointment as arbitrator for the subcontract dispute. The subclause wisely *empowers*, but does not *require*, the arbitrator to order consolidation of the proceedings, or the holding of concurrent hearings.

Subclause 37.5 – via the *special conditions* included in form MF/2 (rev 1) – provides for the arbitration to be conducted in accordance with a named set of rules to be chosen by the parties, and gives as examples the International Chamber of Commerce (ICC), United Nations Commission on International Trade Law (UNCITRAL), London Court of International Arbitration (LCIA), Chartered Institute of Arbitrators (CIArb) or Construction Industry Model Arbitration Rules (CIMAR). Similar arbitration procedures are incorporated in clauses 52 of form MF/1 (rev 3), as supplemented in 1998, and advice is included for arbitration procedures under Scots law.

## 2.8.12 The ICE Arbitration Procedure 1997

The ICE Arbitration Procedure 1997 is an updated version of the Procedure published in 1983, and takes account of the many developments affecting arbitration during the 14-year interval between the two publications. While intended for use with the ICE family of contract forms and with the NEC family, in arbitrations conducted under the Arbitration Act 1996 in England and Wales, the procedure also makes provision for use in other jurisdictions to the extent that the applicable law permits.

Many of the rules contained in the procedure reflect the provisions of the 1996 Act; in some cases in paraphrased form and sometimes with detailed elaboration: for example in the procedure for appointment of the (sole) arbitrator, either by agreement or by the President of the ICE. There are however certain provisions which materially extend the arbitrator's powers.

Rule 5 provides for further disputes to be referred to the arbitrator at any time before his appointment is completed; clarifying the intention that the elaborate preliminaries of clause 66 of the ICE conditions relating to reference of disputes to the engineer, conciliation etc. do not preclude the arbitrator from dealing with all disputes in a single arbitration. Similarly the rule gives the arbitrator jurisdiction over

'any issue connected with and necessary to the determination of any dispute or difference already referred to him whether or not any condition precedent to referring the matter to arbitration had been complied with'.

Rule 9.1 provides a slight extension to the power available to the arbitrator under section 35(1)(b) of the 1996 Act to order concurrent hearings of related disputes. The Act merely empowers the parties to agree upon such consolidation and upon the terms relating thereto; while rule 9.1 empowers an arbitrator to whom all of the related disputes have been referred, upon the application of a party being a party to all of the contracts involved, to *order* such consolidation.

In the common situation of disputes arising from a main contract and from a subcontract for part of the work, consolidation of arbitrations under the two related contracts is usually sought by the main contractor, who is anxious to avoid the possibility of inconsistent awards, both adverse to his interests. The employer and the subcontractor however often oppose consolidation on the ground that it would prolong the proceedings and increase costs: neither of these two parties wishes to become involved in the dispute with which they are not concerned. Under rule 9.1 the arbitrator would be empowered (but not required) to order consolidation, but only if the dissenting parties were subject to the ICE arbitration procedure. Furthermore the arbitrator has an overriding duty, under section 33(1)(b) of the 1996 Act, to avoid unnecessary delay or expense, and should therefore order consolidation against the wishes of two of the three parties involved only if he were confident that it was in their best interests to do so.

Rule 13.7 of the procedure empowers the arbitrator to 'allocate the time available for the hearing between the parties...', implying the use of the 'chess clock' allocation that has become popular in major hearings. Such a power is impliedly available to (and arguably a duty may be imposed upon) the arbitrator under section 33(1)(b) of the 1996 Act, which requires him '...[to] adopt procedures suitable to the circumstances of the particular case, avoiding unnecessary delay or expense...'.

Rule 16.1 which forms part of Part F Short Procedure provides for the arbitrator's fees and expenses to be borne in equal shares between the parties, and removes the arbitrator's power to award costs. The rule is clearly intended, in minor disputes, to discourage the parties from appointing legal or other representation, and to that extent it may sometimes be appropriate. However an undesirable side effect of the rule is that it removes one of the most important powers available to the arbitrator in dealing with defaults and delays caused by one or other of the parties; and it is probably for this reason that such a rule is rendered ineffective by section 60 of the 1996 Act unless entered into after the dispute has arisen.

Rule 19.3 activates section 39 of the 1996 Act, which empowers the arbitrator, where the parties so agree, to make provisional orders giving any relief which the arbitrator would be empowered to grant in his award. Such orders could relate to payment of money or the disposition of property between the parties, and are subject to review in the final award. The purpose of the rule, and of section 39, is to enable the arbitrator to order, at an early stage in the proceedings, payment of money which he is likely to order in his award. Such action is appropriate where, for example, a claimant's impecuniosity may prevent him from pursuing his claims, or where it is clear that the respondent has no defence to a substantial part of the claim. The procedure is available to a claimant as an alternative to adjudication; although it is likely that the latter procedure may be more effective where an early payment is essential to the claimant's financial survival (see Chapter 11 section 11.6).

## 2.8.13 The Construction Industry Model Arbitration Rules

The Construction Industry Model Arbitration Rules (CIMAR 1998) were drafted by the Society of Construction Arbitrators (SCA) after a long period of consultation with interested bodies, and were published in 1998. By that date drafts had been endorsed by most of the leading construction institutions and associations, including the ICE, the ACE, the RIBA, the RICS, the CECA, and the conditions of contract standing joint committee comprising representatives of the ICE, the CECA and the ACE. The rules have been adopted in the ICE conditions of contract seventh edition as an alternative to the ICE's own arbitration procedure, and by the RIBA.

The objective of CIMAR 1998 is set out in rule 1, which encapsulates the provisions of sections 1, 33 and 40 of the 1996 Act: namely

'to provide for the fair, impartial, speedy, cost-effective and binding resolution of construction disputes, with each party having a reasonable opportunity to put his case and to deal with that of his opponent'.

The drafting committee also recognised the need, at a time when construction work spans more than one professional body, for rules to be adopted by all relevant construction institutions and bodies. The record of adoption referred to above indicates the substantial degree of success achieved in that aim.

Early in its deliberations the drafting committee was faced with a question whether to incorporate provisions of the 1996 Act by reference or by extensive repetition. The committee decided, wisely it is submitted, on the former course. In consequence CIMAR 1998, unlike its rival the ICE procedure, is a concise document incorporating a mere 14 rules covering 11 sides of A4 typescript. While the aim of the committee was 'user-friendliness' it also recognised the merits of 'brevity coupled with clarity' which has, the drafting committee state, commanded wide support.

The drafting notes define three purposes of the rules:

- to incorporate powers in the Act;
- to extend or amend provisions of the Act where necessary;
- to add a general framework to the specific powers and duties.

It follows that most of the rules merely restate provisions of the 1996 Act, in a concise form. Certain of the powers given to the arbitrator by the Act are however dependent upon the parties' agreement thereto, and in some cases that agreement is incorporated in CIMAR 1998, in particular the following.

- Under rule 3, provision is made for related disputes arising under the same arbitration agreement to be consolidated with disputes that have already been referred, whether the related disputes are notified before or after the appointment of the arbitrator.
- Under the same rule an arbitrator appointed in two or more related proceedings on the same project, each of which involves some common issue whether or not involving the same parties is empowered to order the concurrent hearing of any two or more such proceedings. This constitutes the agreement needed in order to activate, under certain conditions, section 35(1)(b) of the 1996 Act.
- Under rule 10 the arbitrator is empowered, with minor restrictions, to order provisional relief pursuant to section 39 of the 1996 Act.
- Under rule 11 powers available under section 41 of the 1996 Act in the absence of agreement to the contrary are confirmed. These include the power to dismiss a claim on the ground of inexcusable delay in pursuing it; the power to proceed in the absence of a party or of submissions; and the power under section 41 to deal with other types of default.

It is worthy of note that the power conferred on the arbitrator under rule 3 to order concurrent hearing of disputes does not extend to a power to order consolidation: the rule merely restates the provisions of section 35(1)(a) of the 1996 Act. The omission of such a power from a document approved by a major proportion of the construction industry appears to confirm the wisdom of those who drafted the 1996 Act in not providing for compulsory consolidation.

### 2.8.14 The Chartered Institute of Arbitrators' Arbitration Rules (2000 edition)

The CIArb Arbitration Rules (2000 edition) became effective on 1 December 1999, and incorporate all provisions of the 1996 Act except where a non-mandatory provision of the Act is expressly excluded. The rules also include guidance on matters of procedure relating to the arbitration agreement, the appointment of the arbitrator, and during the interlocutory proceedings.

Rule 7.1 bestows on the arbitrator all of the powers given by the Act, which would of course be available whether or not the CIArb rules applied. However two of the provisions of the Act – consolidation of proceedings under section 35 and provisional orders under section 39 – require the parties' agreement thereto, and such agreement is incorporated in the rule. Rule 7.3 empowers the arbitrator to consolidate, or to order concurrent hearings, where two arbitrations, whether or not involving the same parties, raise common issues of fact or law. The rules also provide for a Short Form Procedure, based on documents only, where the parties so agree.

Under a separate set of rules entitled the Controlled Cost Arbitration Rules (2000 edition) provision is made for the arbitrator to have a duty 'to use his best endeavours to direct the arbitration procedure in such a way that the total costs of the reference (including his own fees and expenses) do not exceed 20% of the sums in issue in the reference.'

# CHAPTER THREE
# APPOINTMENT OF THE ARBITRATOR

## 3.1 Synopsis

A party wishing to refer a dispute to arbitration should, unless the parties concur in the appointment of an arbitrator, comply strictly with the terms of the arbitration agreement relating to appointment procedure and to the qualifications of the arbitrator. Provision is made in the 1996 Act for fall-back procedures where the arbitration agreement is silent or is ineffective, and provision is also made for supplying any vacancy that may arise.

## 3.2 Constitution of the tribunal

The usual practice in the case of arbitrations arising from construction contracts, and especially those of a domestic nature, is to appoint a sole arbitrator, and in the absence of agreement to the contrary, section 15 of the 1996 Act automatically provides for such an appointment. In addition to likely saving in cost, the appointment of a sole arbitrator may help in expediting the arbitration. This is because it reduces the difficulty often encountered in finding suitable dates for meetings and hearings: the existing commitments of the parties, their legal advisers, counsel, and expert witnesses have to be taken into consideration in addition to those of the tribunal, and where the tribunal comprises three or more persons the difficulty is compounded.

Whatever may have been incorporated in the agreement to refer the parties are of course always able to vary that agreement by mutual consent, when a dispute arises or at any other time.

## 3.3 Appointment procedure

Section 18 of the 1996 Act provides a fall-back procedure under which the court is empowered, under subsections (2) and (3), and to the extent that there is no agreement, 'to give directions as to the making of any necessary appointments'. That provision is, however, unlikely to be invoked in the case of construction contracts, because in most cases the arbitration agreement makes valid and suitable provision for appointment of the arbitrator. Many arbitration agreements in construction contracts also provide for attempts to settle disputes by some form of negotiation before arbitration is invoked. The ICE conditions, for example, require that a dispute is initially referred to the engineer for his decision, and it is only if that decision is not acceptable to both parties, or is not given within the specified period, that the next stage, namely conciliation, may be invoked. That stage is however optional, its adoption depending upon the agreement of both parties. Additionally, the parties are entitled, pursuant to the terms of the construction contract or to the Scheme for Construction Contracts (SI 1998 No 649 (see Appendix C)) where the contract terms do not so provide, to invoke adjudication at any time (see Chapter 11 section 11.6).

While the use of some structured form of negotiation, such as mediation, conciliation, or the 'mini-trial', all of which are now referred to as forms of 'alternative dispute resolution' (ADR), is sometimes successful in resolving the dispute, success usually depends upon a genuine willingness on the part of both parties to reach a compromise. Regrettably that willingness is often not common to both parties: a respondent wishing to defer the day upon which payment has to be made will sometimes enter into negotiations or some form of ADR with the sole objective of deferring arbitration. A party anxious to obtain payment of sums considered to be due to it should be aware of the possibility of such tactics, and should invoke arbitration at the earliest stage permitted under the arbitration agreement unless satisfied that its opponent has a genuine desire to compromise the matters in dispute. It is of course always possible to initiate, or to continue, settlement negotiations concurrently with the arbitration.

Many arbitration agreements require a 'notice of arbitration' to be given, defining the date upon which arbitration proceedings are commenced. This may be important in disputes such as those under the ICE Conditions, where unless notice is given within one month of its receipt, the recommendation of a conciliator may be deemed to have been accepted as finally determining the dispute. Similarly the decision of an adjudicator, unless challenged within 3 months by the issue of a 'notice to refer' [to arbitration] also becomes final and binding. The notice of arbitration is often accompanied by a 'notice to concur' (see Appendix A SD/4) specifying the names of, usually, up to three persons nominated as being suitable for the appointment. The recipient of such a notice may agree to one of the named persons, or may reject all of them, possibly nominating up to three others. In some cases it is possible for the parties to reach agreement on the appointment, but where agreement is not reached within 28 days of the notice to concur, (or within one month under the ICE Conditions) either party may apply to the appointing authority named in the agreement, or to the court under section 18 of the 1996 Act, as appropriate, for an appointment to be made (see Appendix A SD/5). Section 14 of the 1996 Act provides that the parties are free to agree when arbitration proceedings are to be regarded as commenced; and in the absence of such agreement proceedings are commenced when one party serves on the other a notice to appoint or to concur in the appointment of an arbitrator; or, where a third party is to make the appointment, when a party requests the third party to make the appointment.

## 3.4 Qualifications of the arbitrator

### 3.4.1 Impartiality

The first and most important qualification of the arbitrator is that he shall be impartial. He must have no interest in or relationship with either party such as might impair his impartiality or be thought to do so; neither must he have any financial interest in either party, for example as a shareholder or a consultant. He should not have any prior knowledge of the subject matter of the dispute, because that knowledge might be incorrect and/or might create difficulty in distinguishing it from the evidence adduced. He should of course have a general knowledge of the type of contract or of construction from which the dispute arises, in order that he may understand the technical issues that may arise.

Where, as sometimes happens, an arbitrator discovers either before or after his appointment that he has some relationship with one of the parties or one of their experts or witnesses, he should where necessary decline the appointment or resign. However if he is satisfied in his own mind that the relationship is irrelevant to the matters he will be called upon to determine, he should disclose the relationship to the parties and be prepared to stand down if either party makes a reasonable objection to his proceeding. This does not mean that he should necessarily accede to an unreasonable request for his replacement where it appears that the primary objective of the challenger is to cause delay, but where reasonable doubt exists he has no alternative but to decline the appointment or to resign as the case may be. The test for impartiality in such cases is whether or not a reasonable person might consider there to be a risk that the arbitrator is not impartial.

### 3.4.2   Technical knowledge

The arbitrator should have a general knowledge of the technicalities of the matters in dispute. The extent to which that knowledge relates to specialised subjects must of course depend upon the variety of the issues that may arise, since it is unlikely that a single arbitrator who has a variety of specialised skills, in addition to a thorough knowledge of arbitration procedure will be found. Usually it is possible to find a competent arbitrator whose construction experience is relevant to the matters in dispute in general terms: for example to roads and bridges, or to building structures, or to water supply, or to whatever type of construction contract led to the dispute. Where specialised knowledge – such as knowledge of some novel technique of construction or of testing – is needed, it may be necessary for the parties to adduce expert evidence on the specialised subjects.

If the arbitrator finds, either before or after his appointment, that the dispute concerns matters of which he has prior knowledge – for example a matter in which he has given specialist advice, or has made a study of for some other purpose, then the extent and details of that prior knowledge should be made known to the parties, and where appropriate the arbitrator should decline the appointment or withdraw from it, as the case may be. This is because of the danger that knowledge gained from his previous involvement may be difficult to distinguish from the evidence adduced by the parties. The arbitrator's duty is to make his decision on, and only on, the evidence adduced at the hearing.

### 3.4.3   Knowledge of arbitration procedure

Of no less importance in ensuring that the arbitration is conducted efficiently and expeditiously is the arbitrator's knowledge of arbitration procedure. He should be able, on learning details of the matters in dispute – usually at the preliminary meeting (see Chapter

5) – to suggest the use of a procedure suited to the content and magnitude of the dispute, to guide the parties towards the formulation of sensible directions on such matters and on the timing of the preliminaries and of the hearing, to conduct the hearing (if any) in a competent and just manner, and to make a just, valid and effective award. The latter requires what is sometimes termed 'judicial capacity' – the ability to consider and weigh the evidence; to determine disputed facts, and to apply a sound knowledge of arbitration law and at least a basic knowledge of contract law, in making a reasoned award. In addition the arbitrator may be called upon to determine issues that arise during the interlocutory proceedings: for example as to whether or not certain documents are privileged, whether or not a party should be permitted to amend its statement of case or be granted an extension of time.

## 3.5  Terms of the arbitrator's appointment

Whether chosen by agreement of the parties or appointed by an authority it is the arbitrator himself who chooses whether or not to accept the appointment. In doing so he may require as a condition of his acceptance that the parties agree to his proposed terms of appointment. In the case of a major arbitration it is desirable, if not essential, that agreement should be reached on terms covering such matters as hourly and/or daily rates of remuneration, provisions for fees in respect of the cancellation of reserved time, initial charges and payments on account, and possibly provision for adjustment of rates to take account of inflation (see, for example, SD/6 and SD/7 in Appendix A). In smaller disputes, where the overall duration of the arbitration is expected to be in months rather than years, it may be sufficient to agree just hourly and daily rates for time 'spent on or allocated to the duties of the reference'; but failure to define charges for cancelled engagements often leads to further disputes.

Where the arbitrator has been appointed by an authority such as the President of the ICE it may be unwise for the arbitrator to make his acceptance conditional upon the agreement of his terms. This is because failure to agree terms is sometimes used by a reluctant party as a delaying tactic, on the basis that the nominee's failure to accept the appointment will result in a further application to the President and hence further delay while a replacement is appointed. It is suggested that the arbitrator should seek the parties' agreement to reasonable terms, but not make his acceptance of the appointment conditional upon such agreement. He can then, if one or both parties fail to agree, rely upon his power under section 63(3) of the 1996 Act which provides, in the absence of agreement to the contrary, that

> 'The tribunal may determine by award the recoverable costs of the arbitration on such basis as it thinks fit.'

Costs of the arbitration are defined in section 59(1)(a) as including the arbitrator's fees and expenses. In addition, the arbitrator is empowered, under section 38(3) to order a claimant to provide security for the costs of the arbitration.

A problem sometimes arises where one party agrees to the arbitrator's proposed charges, and the other party rejects or fails to agree those charges. In *K/S Norjarl* v. *Hyundai Heavy Industries* (1991) 3 WLR 1025 it was held in the High Court that it was inappropriate for an arbitrator to conclude an agreement about fees with one party where the other party did not agree; because by doing so he might lay himself open to an imputation of bias. It follows that where only one party agrees to his proposed terms the arbitrator should proceed as though neither party had agreed, and should notify the parties accordingly.

## 3.6   *Supplying vacancies*

It sometimes happens that after being appointed the arbitrator ceases to be available to conduct the reference. His appointment may be revoked under section 23 of the 1996 Act; or he may be removed by the court under section 24; or he may resign (section 25); or he may die (section 26). All of these situations are dealt with under section 27 of the 1996 Act, which provides, in the absence of agreement to the contrary, for appointments by the same procedures as those relating to the original appointment.

# CHAPTER FOUR
# JURISDICTION OF THE ARBITRATOR

## 4.1 Synopsis

The arbitrator's jurisdiction is derived from the arbitration agreement between the parties, from his appointment by those parties, and from the 1996 Act. Provided that the agreement is valid and that he has been properly appointed, the 1996 Act gives him the powers needed to conduct the proceedings and to make an enforceable award.

## 4.2 Power of arbitrator to rule on own jurisdiction

Challenges to the arbitrator's jurisdiction, based on an allegation of the non-existence of a valid arbitration agreement, on improper constitution of the tribunal, or on the ambit of matters referred, are dealt with in section 30 of the 1996 Act. Unless the parties agree otherwise, the arbitrator is empowered to rule on his own substantive jurisdiction, with the proviso that his ruling may be challenged by any available appeal process in the Act. The court is expressly empowered, under a mandatory provision contained in section 32 of the 1996 Act, to determine any question of the substantive jurisdiction of the arbitrator. It is difficult to envisage any reason why the sensible provisions of section 30 should be varied in an arbitration agreement.

Any challenge relating to an alleged lack of jurisdiction at the outset of the proceedings must be raised by a party not later than the time he takes the first step in the proceedings (section 31 of the 1996 Act). The fact that the party making the challenge may have appointed, or participated in the appointment of the arbitrator does not debar it from making such a challenge. Any objection made during the course of the proceedings must be made as soon as possible after the event giving rise to the objection occurs, with the proviso that the arbitrator may, if he considers the delay to be justified, admit a later objection.

The arbitrator may give his ruling on the objection either in a separate award as to jurisdiction or as part of his award on the merits of the case; with the proviso that he must comply with the parties' wishes where they are in agreement. Where necessary the arbitrator may stay the arbitration proceedings while an application is made to the court under section 32 of the 1996 Act, and he *shall* do so where the parties so agree.

Section 32 provides that the arbitrator's decision as to his jurisdiction may be made the subject of an application to the court for leave to appeal: and that the court, having made its decision, shall not give leave to appeal from that decision unless it considers that the question involves a point of law which is of general importance or which should for some special reason be considered by the Court of Appeal.

## 4.3 Revocation of the arbitrator's authority

Although in earlier enactments the power to revoke an arbitrator's authority was vested only in the High Court or a judge thereof (section 1 of the 1950 Act) it was not unusual for that authority to be revoked by agreement of the parties; for example where the dispute is compromised. More rarely there have been instances in which the parties, having lost confidence in their appointed arbitrator, have agreed to dispense with his services, and have further agreed to continue their arbitration under the jurisdiction of another arbitrator. Provided that the parties honour the terms of their contract with the original arbitrator as to fees and any other matters, such action is unimpeachable.

The 1996 Act recognises the reality of this possible situation in section 23, in which the parties are empowered to agree circumstances in which the authority of their arbitrator may be revoked. In the absence of such agreement, the parties are empowered to revoke the arbitrator's authority by their joint action. The appointing authority may also be empowered by the parties to revoke the arbitrator's authority. In most cases the provisions of the Act are likely to be satisfactory without any special provision in the parties' arbitration agreement.

## 4.4 Power of court to remove the arbitrator

Section 24 of the 1996 Act empowers the court, on the application of a party, to remove an arbitrator whose lack of impartiality, qualifications, physical or mental capacity, or conduct of the proceedings give grounds for such action. The court may so act only after the applicant has exhausted recourse to any authority or person vested with the same power. The arbitrator may, however, continue with the arbitration proceedings and make an award, while such an application is pending. It is submitted that where there is a serious doubt as to whether the arbitrator will remain in office after the application has been heard, the arbitrator should if necessary, and certainly on the application of the parties, suspend the proceedings until the outcome of the application is known.

Where doubts exist as to the arbitrator's impartiality the court is likely to exercise its powers robustly, since the doubt implies a possible infringement of the rules of natural justice: the foundation upon which arbitration is based. In *Turner (East Asia) Pte Ltd* v. *Builders Federal (Hong Kong) Ltd and Josef Gartner & Co* (1988) 42 BLR 122 the applicant for removal of an arbitrator pursuant to section 17 of the Arbitration Ordinance of Singapore had

'amply established that [the arbitrator] might not resolve the question before him with a fair and unprejudiced mind ... [and had shown] ... that in this case there is a real likelihood of bias'.

The application was successful.

## 4.5 Resignation of the arbitrator

Section 25 of the 1996 Act makes provision for the parties to agree with the arbitrator as to the consequences of the arbitrator's resignation: as regards his entitlement to fees and expenses, and any liability he may incur. The fall-back provision, in the absence of agreement, is for an application by the arbitrator to the court, which is empowered to grant relief to the arbitrator and to make such order as it thinks fit.

The circumstances in which an arbitrator might wish to resign, and his reasons for doing so, are in general difficult to foresee at the time of making the arbitration agreement. It is submitted that the provisions of the Act are fair and reasonable, and require no special provision in the arbitration agreement.

## 4.6 Death of the arbitrator or of the appointing authority

The arbitrator's authority is, under section 26 of the 1996 Act, personal, and therefore ceases on his death. Death of the person by whom he was appointed does not affect the arbitrator's authority, unless the parties agree otherwise. Here again no special provision is needed in the arbitration agreement.

## 4.7   Power to conduct the proceedings

### 4.7.1   Procedural and evidential matters

The principal power required by the arbitrator in the interlocutory stages, namely to give directions as to the conduct of the proceedings and all matters relating thereto, is given under section 34 of the 1996 Act, but is subject to the right of the parties to agree any matter. Coupled with that power is a duty imposed on the arbitrator under section 33: namely to

'act fairly and impartially as between the parties, giving each party a reasonable opportunity of putting his case and dealing with that of his opponent'.

That power encapsulates what are often referred to as the rules of natural justice, usually expressed as being that 'no man may be a judge in his own cause' and 'every party has the right to be heard'. Those two rules, and their corollaries, form the foundation of arbitration procedure.

In the absence of the parties' agreement to the contrary, section 34 empowers the arbitrator to determine such matters as the timing, location and language of any part of the proceedings; whether or not there should be an exchange of statements of case (previously termed 'pleadings') and if so the form of such statements; whether or not there should be disclosure of documents (previously termed 'discovery') and at what stage; and whether or not there should be a hearing. Some of these matters are outlined below.

### 4.7.2   Statements of case

Section 34 of the 1996 Act also empowers the arbitrator, subject to the right of the parties to agree any matter, to decide whether or not there should be written statements of case, and if so, their form; when they should be supplied and the extent to which they may later be amended. The term 'pleadings' appears to have been abandoned; possibly because it could be regarded as a term not generally understood by laymen. Pleadings have in recent times gained a reputation for disclosing as little as possible of each party's case, while allegedly complying with the rules under which they are served. If the use of the new terminology results in more straightforward statements, in which the objective is to provide the information needed by each party in preparing its case, then the change of name is to be welcomed.

### 4.7.3  Disclosure of documents

Another important power conferred on the arbitrator by section 34, and again subject to the right of the parties to agree any matter, is the power to decide

'whether any and if so which documents or classes of document should be disclosed between and produced by the parties and if so at what stage'.

Here again the erstwhile term 'discovery' appears to have been abandoned, possibly for the same reason as that in the case of pleadings: but the major innovation in the 1996 Act is the discretion given to the tribunal and to the parties to decide whether or not there should be any form of disclosure at all.

The case against complete disclosure is that it sometimes results, in major arbitrations and in those arising from construction contracts in particular, in vast quantities of documents having to be listed, collated and copied in quadruplicate (one copy for the arbitrator, one for each party, and one for the witness giving evidence), at substantial expenditure both of time and of money. Attempts to distinguish documents that are relevant to the issues in dispute and to confine copying to those documents are rarely successful, possibly because of the cost of the knowledgeable staff needed to identify the relevant documents and because of the danger, real or perceived, that some important document may inadvertently be omitted.

The arguments for complete, or at least selective, discovery are primarily that it results in discovering the truth: and that documentary evidence is generally far more reliable than oral evidence, especially where – as is usual in major construction contract disputes – there has been a considerable lapse of time after completion of a project in bringing the dispute to a hearing.

### 4.7.4 Inquisitorial procedure

Section 34(2)(g) empowers the arbitrator, subject to the right of the parties to agree any matter, to decide 'whether and to what extent the tribunal should take the initiative in ascertaining the facts and the law'. This provision, together with the provision of section 37 of the 1996 Act under which the tribunal is empowered to appoint experts or legal advisers, can surely leave no doubt as to the power of the arbitrator to proceed inquisitorially, subject only to the parties' wishes. Whether or not he should do so is, it is submitted, a decision to be taken in the light of knowledge of the scope and nature of the dispute: it is a matter best dealt with at the preliminary meeting (see Chapter 5, section 5.3).

The adoption of inquisitorial procedure is usually advantageous where the parties do not have legal representation, and are not familiar with arbitration or court procedures. Such parties may have little idea as to how to present their cases, of the issues likely to determine the outcome of the dispute, or of the need to support their allegations by evidence. In such cases the arbitrator may be able, by proceeding inquisitorially, to ascertain the facts from which he is able to make a just award.

### 4.7.5 Powers in case of a party's default

The powers available to the arbitrator in dealing with defaulting parties are contained in sections 39, 40 and 41 of the 1996 Act, and these are, where necessary, supplemented by additional powers of enforcement through the court under section 42.

Under section 39 the parties are free to confer on the arbitrator power to order, on a provisional basis, any relief which he would have power to grant in a final award: but the power is available only where the parties do so confer it upon the arbitrator. Unfortunately this provision is unlikely to be helpful to an arbitrator faced with the common situation where the respondent seeks merely to defer the day on which he will have to pay his debts. Clearly such a respondent would not agree to confer a power under section 39 if it is likely to lead to an order that he makes payment earlier than would otherwise be the case. It is, however, likely that a reluctant party's agreement to conferring the power could be obtained where some form of incentive to doing so is available. For example, where a respondent seeks a prolonged deferment of the hearing date in order that its chosen counsel may be available, it would not be unreasonable for the opposing party to agree to such deferment provided that a provisional order is made for payment of a proportion of the sum claimed. Both the ICE Arbitration Procedure and CIMAR include agreements empowering the arbitrator to grant provisional relief under section 39 (see Chapter 2, subsections 2.8.12 and 2.8.13).

Section 40 imposes a duty on the parties to 'do all things necessary for the proper and expeditious conduct of the arbitral proceedings'; and the duty expressly requires the parties 'to comply without delay with any determination of the tribunal'. Section 41 provides for the dismissal of a claim where there has been 'inordinate and inexcusable delay on the part of the claimant in pursuing his claim'. The section effectively reverses the much criticised decision of the House of Lords in *Bremer Vulkan Schiffbau und Maschinenfabrik* v. *South India Shipping Corporation* [1981] 2 WLR 141. In addition it makes express provision for the arbitrator to proceed *ex parte* in a case where a party defaults in attendance or in complying with an order to submit written evidence or make written submissions. This codifies what has been believed to be a power of the arbitrator, but for which the authority is obscure.

Section 41(6) further provides that where a claimant fails to comply with a peremptory order to provide security for costs the arbitrator may make an award dismissing the claim, and under section 41(7) failure by either party to comply with a peremptory order may result in the arbitrator drawing adverse inferences from the failure and making an award on the basis of such materials as have been properly supplied. Finally, the section empowers the arbitrator

'to make such order as it thinks fit as to the payment of costs of the arbitration incurred in consequence of non-compliance'.

Peremptory orders by the arbitrator are enforceable under section 42 by the court upon the application of the arbitrator or of a party with the permission of the arbitrator. Powers of the arbitrator relating to the award and to costs are dealt with in Chapters 8 and 9.

### 4.7.6 Power to override the engineer's or the architect's decisions

An important power, quite separate from and additional to the arbitrator's powers under the 1996 Act, is contained in the ICE

Conditions of Contract, in the JCT Form of Building Contract, and some of the related forms. The wording used by the ICE Contract is

'... Such arbitrator shall have full power to open up review and revise any decision opinion instruction direction certificate or valuation of the Engineer' or an adjudicator;

while in the JCT Form the words used are

'... the Arbitrator shall ... have power ... to open up, review and revise any certificate, opinion, decision ... requirement or notice and to determine all matters in dispute which shall be submitted to him in the same manner as if no such certificate, opinion, decision, requirement or notice had been given'.

Forms derived from, or related to, the parent forms of civil engineering and building contracts use similar forms of wording.

For many years the courts had construed that additional power (which is unchanged from that in earlier editions of the ICE and JCT conditions) as being available only to an arbitrator appointed pursuant to clause 66 of the ICE conditions, clause 41 of the JCT form, or to corresponding clauses of other forms, and not to the court itself. That construction was consistent with the not unreasonable presumption that the power to overrule the engineer or architect should be exercised only by a person having the technical knowledge and experience in the construction industry necessary for the determination of technical issues.

It originated from *Northern Regional Health Authority* v. *Derek Crouch Construction* (1984) 26 BLR 1, in which the Court of Appeal upheld the judgment of Judge Smout QC. It became known as the *Crouch* principle, and was followed in several important cases during the ensuing 14 years, notably *Crown Estates Commissioners* v. *John Mowlem & Co Ltd* (1994) 70 BLR 1 and *Balfour Beatty Civil Engineering* v. *Docklands Light Railway* (1996) 78 BLR 42. In the *Balfour Beatty* case the contract was based on a modified version of the ICE conditions, fifth edition, from which clause 66 relating to disputes and incorporating the arbitration agreement had been deleted. In consequence there was no express provision for opening up and reviewing the engineer's decisions: indeed there was, by reason of another amendment, no engineer: his functions being exercised by an 'employer's representative' whose title implied even more questionable impartiality than that of an engineer appointed and paid by the employer. Upholding the judgment of Judge Cyril Newman QC who had held that the court had no general power to open up, review and revise decisions of the employer, Sir Thomas Bingham MR commented in the Court of Appeal that 'It is not for the court to decide whether the contractor made a good bargain or a bad one; it can only give effect to what the parties agreed'.

However the decision in *Crouch* and in subsequent cases decided on the same principle was overruled by the House of Lords in *Beaufort Developments (NI) Ltd* v. *Gilbert Ash NI Ltd* (1998) 88 BLR 1, in which the Court of Appeal of Northern Ireland had found itself constrained by the English Court of Appeal's decision in *Crouch*, even though it considered that decision to be wrong. The House of Lords confirmed that the court has inherent power to 'open up, review and revise' any certificate etc. of the architect.

That decision, which arose from a building contract under the JCT form but is also relevant to contracts under the ICE and other forms, was based upon the court's finding that a certificate of the architect was not, on the true construction of the contract, final and binding, as had been held in *Crouch*.

In his leading judgment Lord Hoffman explained the basis upon which he rejected the notion that decisions of the architect were final and binding with the words

'The judgments of the Court of Appeal contain no very detailed analysis of the provisions of the contract which are said to confer upon the architect this power to issue binding certificates. Although none of the judges say so expressly, there is an implied suggestion that one can infer such power from the very fact that the arbitrator is given a power to "open up, review and revise". This is the argument from redundancy; the parties are presumed not to say anything unnecessarily and unless the decisions of the architect were binding, there would be no need to confer upon the arbitrator an express power to open up, review and revise them. . . .

I think my lords that the argument from redundancy is seldom an entirely secure one . . . even in legal documents . . . people often use superfluous words.'

Later in his judgment Lord Hoffman referred to a much earlier decision upon the same issue, namely the Court of Appeal's judgment in *Robins* v. *Goddard* [1905] KB 294, when the 'open up, review and revise' formula was relatively new, holding that

'... the draftsman had seen no need to confer an express power on the court in the same terms as the arbitration clause. The court's jurisdiction was unlimited. It was the arbitrator's powers which need to be spelled out. On this view, the power to open up, review and revise falls into place alongside the other powers conferred by clause 41.4 as a power which a court would in any event possess.'

Their lordships recognised however that it would be open to the parties to a construction contract to use a form of words which had the effect of making certificates of the architect final and binding, and Lord Hoffman illustrated that freedom with the words

'The contract may say that the value of the property or the question of whether the goods comply with the description shall be determined by a named expert. In such a case, the agreement is to sell at what the expert considers to be the value or to buy goods which the expert considers to be in accordance with the description. The court's view on these questions is irrelevant.'

It is therefore open to those responsible for drafting standard forms of construction industry contracts to revise the present forms so as to give effect to the presumed intention, namely that issues requiring technical knowledge for their determination should be determined only by a person, such as an arbitrator or an expert, who has that technical knowledge. Provision of this type is made in the Institution of Chemical Engineers' Model Form of Conditions of Contract for Process Plant 'the Red Book': Fourth Edition (2001), which will provide for reference of disputes for determination by an expert (see subsection 2.8.10 above).

# CHAPTER FIVE
# THE PRELIMINARIES

## 5.1  Synopsis

Before matters in dispute can be brought to a hearing or to written submissions the issues to be determined must be defined. Each party must be forewarned of the case it has to answer: and documentary evidence must, where the parties so agree or the arbitrator so directs, be disclosed and collated for exchange between the parties and for submission to the arbitrator. Where there is to be oral evidence arrangements must be made for the hearing, while if the dispute is to be determined on documentary evidence only a procedure must be formulated for its submission. A timetable for all of these matters must be made, by agreement of the parties or by direction of the arbitrator.

## 5.2  English law

It is a basic principle of English law that surprise tactics shall not be used, either in litigation or in arbitration. Neither party is allowed to gain an advantage by, for example, making some allegation that has not been disclosed to the other party and for which the other party has been unable to prepare, or by appointing leading or junior counsel to represent them at the hearing without giving notice to that effect. Neither party may without notice call an expert to give evidence, because without such notice the other party would be unable to prepare, possibly by calling its own expert to refute the evidence of the opposing expert. Under section 34(2)(d) of the 1996 Act the extent to which documents are to be disclosed between the parties, and at what stage, are matters within the arbitrator's discretion, subject to the right of the parties to agree on those matters. All of these matters need to be addressed and dealt with, preferably by agreement, but otherwise by direction of the arbitrator. And before giving any such directions the arbitrator must give the parties an opportunity to express their views.

In all but the smallest and simplest arbitrations it is desirable that the arbitrator should convene a preliminary meeting for the purpose of hearing the parties on the above and on other procedural matters that usually arise; for formulating directions for the conduct of the arbitration, and a timetable for the various procedural steps. Consideration can be given at such a meeting to the most economical and expeditious manner in which the arbitration should be conducted, and which of the many procedural options available under the 1996 Act should be adopted.

In some cases the parties may suggest to the arbitrator that they formulate their own directions as to procedure without the need for a meeting, thereby saving the cost of the meeting. While this may in some cases be appropriate, there are many important benefits to be gained from a meeting. The 1996 Act has introduced a number of procedural options aimed at promoting the basic aims referred to in section 1 of the Act: namely fairness, expedition and economy. The benefits to be gained from the adoption of such options, either by agreement of the parties or, where the parties are not in agreement, by direction of the arbitrator, may be lost unless they are understood by the parties. The arbitrator's own knowledge of those options should be of value in determining the most suitable procedure in the light of the type, magnitude and content of the dispute.

Where it is clear to the arbitrator from the documents available to him at the time of his appointment that the dispute is of a very minor nature – involving say a sum of less than five figures at 2000 prices – he may suggest the adoption of a procedure designed to keep costs commensurate with the sum in dispute (see section 5.15). In most other cases, however, the benefits to be gained by convening a preliminary meeting far outweigh the small costs involved.

## 5.3 *The preliminary meeting*

The arbitrator usually convenes a preliminary meeting, or a 'meeting for directions', at which he expects or requires both parties and/or their legal or other representatives to be present. The date, time and venue for such a meeting should as far as possible be chosen to suit both parties, and the arbitrator should provide the parties with an agenda (see, for example, SD/8 in Appendix A) in order that they may be aware of the matters to be discussed and the decisions to be taken. Some arbitrators still persist in referring to

56

that document as a 'check list', impliedly for their own reference and not for that of the parties. It is however submitted that for every other type of meeting those attending are forewarned of the matters to be discussed by the provision of an agenda; and that the parties attending a preliminary meeting in an arbitration are entitled to similar courtesy.

Having taken a note of the names and appointments of those in attendance, the arbitrator should address, and invite comments on, some or all of the following matters:

### 5.3.1 The contract

If he has not already been provided with a copy of the contract from which the dispute arises the arbitrator should request that he be provided with such a copy for his retention, and should ensure that the arbitration agreement therein is valid and that he has been validly appointed. Should there be any doubt as to those matters his authority may be confirmed by requiring the parties to enter into an *ad hoc* arbitration agreement (see, for example, SD/1 in Appendix A), naming him as the appointed arbitrator. It may also be necessary for the arbitrator to obtain the parties' agreement to his proposed terms of appointment if they have not already indicated that agreement.

### 5.3.2 Outline of dispute

In some cases the arbitrator may have been provided with a brief indication of the matters in dispute; but in many cases there is a need to ascertain and determine, in outline, the nature and amount of the claim or claims; whether the issues to be determined are of fact or of law or both; whether there is considered to be a need for expert evidence; and the likely duration of the preliminaries and of the hearing or hearings. Such an outline is usually provided by the claimant, but the arbitrator should also invite the respondent to comment. In addition the respondent should be invited to indicate whether or not there will be a counterclaim, and if so its nature and magnitude.

### 5.3.3 Procedural matters

SD/8 (in Appendix A) provides a typical list of procedural matters that may need to be addressed at a preliminary meeting. That list is not intended to be exhaustive or to be universally applicable; it covers most of the matters that often arise, but needs to be edited to suit the particular arbitration for which it is to be issued. For example, the item relating to languages and to translations applies only where one or more of the parties or witnesses does not speak English, and the items relating to consolidation and concurrent hearings are relevant only in cases where there are related disputes, such as those arising from subcontracts where the primary dispute is between the employer and the main contractor.

## Venues for meetings and hearings

Venues for meetings and hearings should be chosen, preferably by agreement of the parties, to suit their convenience. Where the parties are not in agreement the arbitrator must make a decision, which should take into account the locations of the parties, their witnesses, and their legal or other representatives. A venue located mid-way between the parties' bases is not necessarily the preferred option where disagreement arises, because it may involve the need to hire premises which might be freely available at one of the parties', or their advisers', headquarters. However parties may sometimes object, not unreasonably, to the use of the opposing party's premises, which may be seen as giving a tactical advantage to that party in that it has facilities and personnel readily available should a need arise. Where such a venue is proposed by a party, it should be adopted only with the agreement of the other party.

## Written statements of case

Written statements of case, or 'pleadings' in earlier terminology, are essential in one form or another in order that the claim, defence, and the matters to be alleged may be defined, forewarning both parties of the cases they have to answer, and that from such definitions the issues may be defined. Although section 34(2)(c) of the 1996 Act implies that the use of written statements of case is optional it is difficult to envisage any arbitration arising from a construction contract being conducted without the use of such statements. In the unlikely event that the parties agree not to use written statements of case the arbitrator should seek to persuade them of the need for such documents and of the difficulties likely to result from their agreement, and should if necessary convene a meeting for the purpose of recording the basis of each party's case. A timetable for submission of the several statements should be defined, preferably by agreement. (See section 5.5).

## Disclosure of documents

The second main stage in the preliminaries is termed, under the 1996 Act, 'disclosure of documents', a procedure formerly termed 'discovery'. Section 34(1)(d) empowers the arbitrator, subject to the right of the parties to agree any matter, to decide 'whether and if so which documents or classes of document should be disclosed between and produced by the parties and at what stage'. Here again it is difficult to envisage a construction contract arbitration in which at least a limited form of disclosure is not adopted. While it is to be hoped that full discovery, involving the copying in quadruplicate of every document related in any way to the contract and to the work performed under it, will not be adopted, it is submitted that all documents that are relevant to the matters in dispute should be made available at the hearing (see section 5.9). Each party should also be entitled to inspect its opponent's documents, excluding those that are privileged (see Chapter 6, section 6.5). Should there be disagreement as to which documents fall into that category the arbitrator must decide, after having heard both parties' submissions on that question. Following upon inspection the parties should each prepare a list of those documents upon which they rely; and it is suggested that a list combining both parties' lists should be agreed as being the relevant documents to be adduced in evidence. Copies should be made of those, and only those, documents for incorporation in agreed bundles.

## Application of the strict rules of evidence

Whether or not the strict rules of evidence should apply to the arbitration is now, under section 34(2)(f) of the 1996 Act, a matter for the arbitrator to decide, subject to the right of the parties to agree that question. Where the parties agree, or in the absence of agreement the arbitrator directs, that the rules of evidence shall apply, the statute defining those rules is the Civil Evidence Act 1995, which repeals corresponding parts of the Acts of 1968 and 1972, and which came into force in January 1997. The principal effect of a decision that the strict rules of evidence shall apply is that where hearsay evidence is sought to be adduced by a party, notice and, where called for, particulars of that evidence must be given. The Act also empowers the tribunal to estimate the weight to be given to hearsay evidence, having regard to the relevant circumstances. In many construction arbitrations there is no need for either party to rely on hearsay evidence, but where such a need arises and the parties are not in agreement as to the basic question whether or not to apply the strict rules of evidence, their submissions should be heard before the arbitrator makes his decision.

## Inquisitorial procedure

Section 34(2)(g) of the 1996 Act provides that the arbitrator may, subject to the right of the parties to agree otherwise, take the initiative in ascertaining the facts or the law. That provision is likely to be of value where the scale of the dispute is so small as not to justify the appointment of any form of representation of the parties – legal or technical. Lay persons attempting to become their own advocates are often quite incapable of presenting their cases and of challenging the opposing case, leaving the arbitrator in ignorance of the facts and submissions he requires to know in reaching a just decision. By using his own initiative in such circumstances an experienced arbitrator can ascertain the facts he needs to know, and the law he needs to apply, in making a just award.

## The choice between written or oral submissions

The choice between written or oral submissions, which is, under section 34(2)(h) of the 1996 Act, at the discretion of the arbitrator subject to the right of the parties to agree, is in general influenced mainly by the magnitude of the dispute. In any but the smallest disputes it is traditional to adopt oral proceedings; but these have in recent times been supplemented by proofs of evidence of witnesses, and written opening and closing addresses by counsel. These innovations have generally enabled the duration of the hearing to be kept to a minimum, with consequent savings in time and in costs. For small disputes 'documents only' procedures are often the more appropriate and economical option: see section 5.15.

## Consolidation of arbitration proceedings

The options available under section 35 of the 1996 Act, where the parties so agree; to consolidate arbitration proceedings with other arbitral proceedings, and/or to hold concurrent hearings, may appear attractive in many employer/contractor/subcontractor disputes, but the practical difficulties often outweigh the advantages. The dispute under the subcontract may form only a small part of the main contract dispute, and the subcontractor may with some justification object to having to await the outcome of matters unrelated to his interests. The CECA Form of Subcontract, in common with some other standard forms, makes provision for 'disputes under the subcontract to be dealt jointly with the dispute under the main contract and in a like manner'. This provision usually results at least in the same arbitrator being appointed to deal with both disputes: but the extent to which the proceedings can be consolidated depends upon the content of the two or more related disputes, and upon the agreement of all of the parties involved to the type of consolidation proposed. In *Redland Aggregates Ltd* v. *Shepherd Hill Civil Engineering Ltd* [1999] BLR 252, it was held in the Court of Appeal that the operation of clause 18(2) of the FCEC 'blue form' did not entitle the main contractor (Shepherd Hill) to arrange a tripartite arbitration (if they could) to suit their own convenience and no other. Sir Christopher Staughton, giving the judgment of the court, stated

'... in so far as the progress of a tripartite arbitration depends on the contractors, I consider that it ought to be set up and conducted with all deliberate speed.'

Allowing an appeal from Mr Recorder Knight QC, Sir Christopher stated

'... the plaintiffs are no longer obliged to take part in a tripartite arbitration, and may call upon the President of the Institution of Civil Engineers to appoint an arbitrator on their disputes with the defendant'.

Earlier in his judgment Sir Christopher had stated

'English law contains no provision for compulsory third party proceedings in arbitration, which is a regrettable lapse for building contractors. Nor is it remedied by the Arbitration Act 1996, which merely states in section 35 that the parties may agree to consolidation of disputes, or may agree to confer on a tribunal the power to order consolidation.'

His judgment in *Redland Aggregates* however provides a clear example of a reason why compulsory consolidation was *not* provided for in the 1996 Act. Such a provision would in this case have resulted in the very mischief that Sir Christopher's judgment corrected. Lord Saville deals with this point on page x in his foreword to this book.

The judgment was upheld on appeal to the House of Lords (*The Times*, 11 August 2000) in which it was held, by a majority, that clause 66 of the ICE conditions envisaged concurrent hearings under rule 7 of the ICE arbitration procedure (1983). If that could not be or was not done, the contractor would no longer be able to use clause 18(2) and the subcontractor would be free to pursue a separate arbitration under clause 18(1).

## Choice of representation

The parties' intentions as to their representation, referred to in section 36 of the 1996 Act, should be ascertained at the preliminary meeting and confirmed in 'Procedural Directions' (see, for example, SD/9 in Appendix A) so that each party may be aware of its opponent's intention. In general it is desirable that the status of each party's advocate (solicitor, junior counsel or leading counsel) should be the same; and arbitrators may sometimes suggest an appropriate level of such representation having regard to the nature and scale of the dispute. In the absence of agreement however, the parties remain free to make their own decision, subject to their giving notice of their intention. The arbitrator may in some cases indicate that he will need to be persuaded, subject to the provisions of section 63 of the 1996 Act, that costs of a proposed level of representation have been incurred necessarily; and such an indication may be a powerful means of persuading parties not to appoint representation at a higher level than is necessary, having regard to the magnitude and nature of the dispute.

## The appointment of experts

The appointment of experts may also need to be considered where the parties or the arbitrator so suggest. Expert opinion on a technical issue may be necessary, except where that issue falls within the arbitrator's own expertise. Experts on quantum, on the other hand, are often of considerable benefit to both parties where they are able to report on their own party's detailed evaluation of the claims and thereafter to meet in order to agree figures as figures so far as may be possible. Where such agreement is reached – usually subject to the resolution of the identified issues – a joint report by the experts defining the values of the claims, taking account of alternative findings on the various issues, is often of great value in reducing the duration of the hearing.

*Security for costs*

The arbitrator may under section 38(3) of the 1996 Act, except where the parties agree otherwise, order a claimant to provide security for costs of the arbitration; i.e. for the costs both of the respondent and of the arbitrator himself. The power to order security for the respondent's costs should be exercised only upon the application of the respondent and after hearing both parties' submissions on that matter, and the limitation contained in section 38(3) of the 1996 Act should of course be observed. Security for the arbitrator's costs is normally dealt with in the arbitrator's terms of appointment, but where either or both parties have refused to agree those terms then the arbitrator's power to secure his own costs is available.

The power to order a claimant to provide security for costs does of course apply equally to a counterclaimant, where the counterclaim is independent of the claim, i.e. where it could be pursued separately if the claim were withdrawn. Other than the two negative provisions of section 38(3)(a) and (b) the Act gives no guidance as to the way in which the arbitrator should exercise his discretion in deciding whether or not to grant an application for security of a respondent's costs. It is reasonable to infer that the absence of positive guidance in the 1996 Act is intended to give the arbitrator wide discretion as to the exercise of his powers under section 38(3), subject only to his complying with the general provisions of sections 1 and 33, relating principally to fairness and impartiality. It is however, in the absence of statutory authority or precedent, suggested that the arbitrator might, when considering an application under section 38(3), have regard to some or all of the following considerations.

- Is the claimant a limited company and therefore able or likely to go into liquidation?
- Would the assets of a parent company be available in the event of liquidation of the claimant?
- Is there an enforceable guarantee of a director of the limited company available in the event of liquidation?
- Is any other form of security available e.g. the guarantee of a bank or an insurance company?

- Is the respondent's fear that the claimant might be unable to pay costs justified?
- Is the respondent's application made in order to oppress the claimant?
- Is the claimant's impecuniosity brought about by the respondent's failure to pay alleged debts forming the subject of the dispute?
- Has the respondent offered a sum in settlement, and if so what is its amount?

The final item in the above list relates to a matter of which the arbitrator would in general be careful to ensure that he has no knowledge, on the principle that such knowledge might prejudice the arbitrator's impartial assessment of the claim on its merits. Ideally this difficulty could be overcome where arrangements can be made, by agreement, for another arbitrator to deal with and determine the application for security of costs. Failing such an arrangement the arbitrator would have to undertake the task himself, and in due course to erase from his mind his knowledge of the offer when he came to deal with the substantive issues.

Where an order for security for costs is not complied with the arbitrator has the power under section 41(5) of the 1996 Act to make a peremptory order to the same effect, requiring compliance within a specified period. Should the claimant fail to comply with that order, the arbitrator is empowered under section 41(6) of the Act to make an award dismissing the claim.

*Evidence on oath or affirmation*

Whether or not evidence should be taken on oath or on affirmation is a matter within the arbitrator's discretion subject to agreement of the parties, under section 38(5) of the 1996 Act. Where the parties are not in agreement it is suggested that the arbitrator should exercise his discretion in favour of requiring evidence to be given on oath or affirmation.

*Directions for the preservation of evidence*

The arbitrator's power, under section 38(6) of the 1996 Act, subject to the parties' agreement to the contrary, to give directions for the preservation of evidence, may sometimes be of value where there is a risk that a party may destroy documents or other evidence, perhaps because its evidential value is not realised. The section could also be applied to real evidence, such as allegedly defective work; and while it may not be practical to preserve the whole of a defective structure, samples of the materials used in its construction could be preserved; or where necessary a written report and photographs of the structure could be obtained and agreed between the parties. It is to be expected that an application for an order under the section will be made by a party aware of the likelihood that evidence may be destroyed.

### Power to award provisional relief

Because of the need for both parties' agreement thereto it is unlikely that the power conferred under section 39 of the 1996 Act to order provisional relief will be available, except where the parties have so agreed before the dispute arises. Such an agreement is incorporated in the current ICE Arbitration Procedure and in the Construction Industry Model Arbitration Rules (see Chapter 2, subsection 2.8.13). Alternatively the Construction Act 1996 has introduced a means whereby such a party may be able to obtain such relief by way of adjudication under the contract or, where the contract does not so provide, under the Scheme for Construction Contracts (see Chapter 11, section 11.6). A claim under the adjudication provisions of the contract or of the Scheme, as the case may be, may be pursued at any time, and may therefore proceed concurrently with an arbitration.

### Exclusion agreement

Section 45 of the 1996 Act enables the parties to enter into an 'exclusion agreement' excluding the jurisdiction of the court to determine questions of law that may arise during the arbitral proceedings. Similar provision is made in section 69 for excluding the jurisdiction of the court to determine a question of law arising from the award. A provision made under section 87 of the 1996 Act, for exclusion agreements to be ineffective in the case of a domestic dispute (unless entered into after the dispute has arisen) has not been brought into operation, and is likely to be repealed. The parties are well advised to enter into such an agreement where their principal aim is finality and where they are content to accept the arbitrator's decision on questions of law as being final. An agreement to dispense with reasons in the award is deemed to constitute an exclusion agreement.

*Separate awards to cover separate issues*

Another important matter for discussion at the preliminary meeting is the use of separate awards to cover separate issues. In earlier legislation such awards were termed 'interim awards', but the 1996 Act has eschewed that term on the ground that it is confusing, and arguably a misnomer (see Chapter 8, subsection 8.4.1). It is often desirable that large and complex disputes be divided into manageable sections, possibly covering different phases of a major contract, or separating issues of liability from those of quantum. Again, it is often desirable that the arbitrator's award of costs is dealt with separately from the substantive issues, so that the arbitrator may hear and take into account the parties' submissions on costs after making his award on the substantive issues.

*Forms of remedy*

Section 48 of the 1996 Act deals with the various forms of remedy available to the claimant, each of which is reflected in the type of award, which may be declaratory, monetary, of specific performance, or of rectification. The claimant should state at the preliminary meeting the type of relief he is seeking; and the arbitrator is empowered, in the absence of agreement between the parties, to determine the appropriate type of remedy to be considered.

*Simple or compound interest*

Section 49 of the 1996 Act has expanded the power of the arbitrator, subject to the parties' agreement to the contrary, to award either simple or compound interest. It is to be expected that in most cases the claimant will seek compound interest, as being the basis on which interest is either paid or earned in commerce and hence a more equitable basis of compensation for being held out of his money.

*Power to limit recoverable costs*

Consideration may be given to section 65 of the 1996 Act, under which the arbitrator is empowered, unless the parties agree otherwise, to limit recoverable costs in the arbitration. Such a limit may be made or varied at any stage, but it must be done sufficiently in advance of incurring the costs to which it relates for the limit to be taken into account.

## Timetable for the preliminaries

Having determined the above procedural matters, preferably by agreement but where necessary by direction of the arbitrator, it becomes necessary to determine a timetable for the various stages of the preliminaries, the first stage of which is usually the exchange of statements of case. The periods of time to be allowed for each stage should where possible be agreed, but where it appears that a party is seeking an unnecessarily long period of time – possibly as a means of delaying the proceedings – the arbitrator may have to impose a limit.

## Disclosure of documents

Secondly arrangements are needed, where it has been decided to adopt some form of disclosure of documents, for each party to inspect its opponent's documents, noting those to be adduced in evidence, and for the preparation of agreed bundles.

## Exchange of proofs of evidence etc.

Dates must also be determined for the final stages of preparation for the hearing, namely the exchange of proofs of evidence of witnesses of fact, for the exchange of reports of experts, where appointed, and for the preparation of agreed reports; and for counsel's opening submissions.

## Arrangements for the hearing

Finally, arrangements for the hearing need to be made, again preferably by agreement, but where necessary by direction of the arbitrator. Such arrangements include dates, sitting hours, venue, and where necessary a transcript of all or part of the proceedings. In addition the parties may wish to agree some form of limitation of the time available to each party for the presentation of its case. The 'chess clock' procedure, under which a period of time for the whole hearing is agreed between the parties and divided equally between them, sometimes provides an effective means of limiting the costs of the hearing, usually by far the most expensive part of the proceedings. Time is charged against each party whenever that party's counsel is on his feet: that is, when giving opening and closing addresses, examining-in-chief and re-examining his own witnesses, and cross-examining the opposing party's witnesses. The procedure does of course require a record to be maintained of the time expended by each party; but that is usually done by a junior member of one of the solicitors' staffs.

## 5.4  Procedural directions

As soon as possible after the preliminary meeting the arbitrator should issue his Procedural Directions, which is a document setting down the arrangements agreed or directed at the preliminary meeting and in addition other standard directions (see, for example, SD/9 in Appendix A).

## 5.5  Statements of case

Each of the several documents in the exchange of statements, previously known as pleadings, should state the material facts upon which each party will rely, not the evidence by which such facts are to be proved. In addition to those facts the statements may include submissions as to the law applicable to the facts. Each statement must include a *statement of truth* signed by the party or by the party's legal representative (CPR1998 rule 22.1). The purpose of the exchange is to forewarn each party of the case it has to answer and to define the matters in issue between the parties. The sequence of the statements is:

- statement of claim;
- statement of defence;
- statement of reply.

Where there is a counterclaim it is pleaded in the same sequence but one stage behind the claim, so that the exchange comprises:

- statement of claim;
- statement of defence and counterclaim;
- statement of reply and defence to counterclaim;
- reply to defence to counterclaim.
  (See SD/10, /11 and /12 in Appendix A).

### Statement of claim

The statement of claim should preferably be in the form of a series of numbered paragraphs, each covering a single allegation. The statement usually commences with a brief introduction of the parties and of the contract between them, drawing particular attention to clauses in the contract on which the claimant relies. This is followed by an account of the events that gave rise, in the claimant's opinion, to the claimant's entitlement to additional payment or whatever other relief is sought. The claimant must define the relief sought in respect of each item of the claim. In most cases that relief is in the form of a sum of money, but there are occasions on which the claimant may seek a declaration, or performance of certain specified work, or rectification of some clause in a contract.

### Statement of defence

Following upon receipt of the statement of claim the respondent is required to submit his statement of defence. Where the claim is in the form of numbered paragraphs, the defence may often deal quite briefly with each of those paragraphs, stating whether its content is 'admitted', or 'denied', or 'not admitted'. The significance of the third statement is that, while the respondent is unable to deny the claimant's statement, he requires the claimant to prove that it is true. In addition to traversing all of the claimant's allegations the respondent may introduce other allegations, such as allocating the cause of delay not, as may be alleged by the claimant, to variations, but to the claimant having insufficient plant and labour on the site. Where a statement in the claim is neither denied nor expressly not admitted it is deemed to have been admitted. Consequently the party drafting the statement of defence should be careful to ensure that all allegations in the statement of claim are adequately covered; and as an added precaution it is usual to add a blanket denial such as

'Save as is herein expressly admitted the Respondent denies each and every allegation contained in the Statement of Claim as if the same were set out and expressly denied *seriatim*'.

### Statement of reply

The third stage, termed the statement of reply, may be required in order to give the claimant an opportunity to deal with any allegations in the defence that are not mere rebuttals of the claim. Any such reply – and replies are not always necessary – must deal only with fresh allegations in the defence: it must not raise any new allegations, since to do so would deny the respondent an opportunity to reply to those allegations.

### Categories of claim

In general there are two main categories of claims: those that arise from clauses of the contract giving an entitlement to extra payment, and those that arise from allegations of breaches of contract. Claims in the former category are evaluated in accordance with terms of the contract: for example Clause 52 of the ICE Conditions, which defines principles to be used in evaluating varied work. Claims for breach of contract are based on allegations of failure to comply with terms of the contract: for example a specified date upon which the contractor is to be given access to the site in order to commence work. Such claims fall to be evaluated as 'damages', in accordance with the principle that the wronged party should as nearly as possible be put in the position he would have been in had there been no breach. In many cases it is in the claimant's interests to plead his claims as alternatives, basing the claim either on an allegation of breach of contract or, in the alternative, as giving an entitlement to payment under a term of the contract.

## 5.6  Amendments to statements of case

It sometimes becomes necessary for a party to seek leave to amend its statement of case. Such a need may arise at any stage from its initial submission until the hearing, and even during the hearing; sometimes from allegations of the opposing party; from disclosure of documents; where fresh evidence comes to light from some other source; or where, during the hearing, a party submits that matters being alleged by its opponent were not disclosed during the pleadings. The party seeking leave to amend must apply to the arbitrator for that leave, giving reasons for its application and including a draft of the amendment proposed, and of course sending a copy of the application and supporting documents to the other party. Where leave to amend is opposed by the other party it may be necessary for the arbitrator to convene an interlocutory meeting in order to hear both parties' submissions before making his decision.

Where leave to amend is given the usual provisos are that the opposing party has leave to make consequential amendments to its pleadings, and that all costs resulting from the amendment and its consequences are to be borne by the applicant in any event, that is, irrespective of the outcome of the arbitration. There are however exceptions to that general rule. Where, for example, the need to amend arises from late disclosure of a document, or some other fault of the opposing party, the applicant should draw such facts to the attention of the arbitrator and seek an order for costs in its own favour.

The consequences of an amendment to pleadings may vary widely depending upon the amendment and on the stage at which it is made. During the early stages of exchange of statements of case the amendment can often be accommodated without difficulty or delay; whereas an amendment made shortly before or during the hearing may necessitate adjournment of the hearing and substantial costs being thrown away in vacating time reserved by the arbitrator and by counsel and others.

## 5.7   *Further and better particulars*

Where a party finds that its opponent's statement is insufficiently detailed to provide an adequate indication of its case, the procedure used under the 1950 Act, and in litigation, has been to request 'further and better particulars' of the statement in question. The 1996 Act leaves it to the arbitrator, subject to the right of the parties to agree, to decide '... the extent to which such statements can later be amended', implying that the parties, or the arbitrator, may decide not to allow applications for amendment or for further and better particulars. While the laudable aim of such a ruling is that the original statement should be sufficiently detailed not to require further elaboration, such an aim is not always achieved. For example, a statement of claim including the assertion that 'the work was delayed and disrupted by the issue of numerous variation orders' clearly requires some definition of the variation orders referred to, without which the opposing party might be unable to prepare his defence. Hence a question might arise as to whether the statement provides adequate warning of the case to be answered, and whether the arbitrator would be acting unfairly if he failed to order the delivery of the particulars requested.

## 5.8  *The Scott schedule*

Where a claim comprises a large number of items, each of which has a separate basis in the contract, it is often convenient for one of the parties – usually the claimant but sometimes both parties – to prepare a schedule in summary form, listing briefly in respect of each item the basis of the claim, the amount claimed, the defence, and the amount (if any) of the respondent's evaluation. See, for example, SD/13 in Appendix A.

## 5.9  *Disclosure of documents*

It is difficult to envisage dispensing entirely with disclosure of documents in any substantial arbitration arising from a construction contract dispute. Many such disputes arise from contracts in which the contract documents comprise conditions, specifications, bills of quantities, site investigation reports, drawings and possibly other documents. Work may have extended over many months if not years, during which there has been voluminous correspondence between the parties and others; several of the site staff may have maintained detailed site diaries; and there may have been numerous site instructions, variations, valuations, interim certificates, notices of claims and other documents, which together form a formidable quantity of records. Witnesses of fact cannot be expected to remember details of the events that occurred possibly years earlier during the progress of the works, hence the only reliable record is that contained in the documents. It is therefore submitted that the parties should dispense with disclosure of documents only if they are content that the arbitrator should make his decision taking into account only the comparatively unreliable oral evidence of witnesses, and ignoring the truth as contained in the written records. Arguments for and against disclosure of documents are referred to in Chapter 4, subsection 4.7.3.

## 5.10  *Privilege*

Certain types of document are privileged from disclosure in evidence. They must however be included in lists of documents, where the production of such lists is ordered, in a separate schedule marked to indicate that privilege is claimed. In construction contract disputes the two main categories of document for which privilege is usually claimed are professional confidences, and offers to settle claims, where marked 'without prejudice' or 'without prejudice save as to costs' (see Chapter 6, section 6.5).

It is sometimes claimed that certain documents should be privileged on the ground that they contain confidential information. Such claims can rarely be upheld, first because arbitration proceedings are private, so that confidentiality is generally preserved, and second because the requirements of justice must take precedence over those relating to confidentiality. In *Mitchell Construction Kinnear Moodie Group* v. *East Anglia Regional Hospital Board* [1971] CLY 375, personal files relating to the contractor's employees were ordered to be disclosed, it being held that the sole issue was one of relevance. A similar principle would apply to the disclosure of other private documents, internal memoranda, pricing and estimating notes etc., where they can be shown to be relevant to the matters in issue.

## 5.11 *Transcript of the hearing*

Although modern electronic methods of producing a transcript of the hearing – sometimes with additional facilities such as provision for its annotation during the hearing – have gained in popularity in the 1990s, the cost of a transcript remains a substantial consideration. The arbitrator should not arrange for a transcript to be made without the agreement of the parties, although he may perhaps suggest their so agreeing in the case of a major dispute: such as one involving a sum running into seven figures or more at 2000 prices. The advantages to be gained from the use of a transcript are first that it leaves the arbitrator free to concentrate on the evidence being adduced, without having to concern himself with making a written record; second that it enables evidence to be taken at a much greater speed than when counsel and witnesses have to wait while the arbitrator completes his note of every question and of every answer; and third that it provides a full and reliable record of the oral evidence and submissions.

Where the scale of the dispute does not so warrant, or where the parties do not agree to the production of a transcript, a tape recording of the proceedings is often of value in supplementing the arbitrator's notes. By keeping a record of the tape number and meter reading at important events – such as the start of each witness's cross-examination – a valuable and accessible record can be made available for later reference, at negligible cost.

## 5.12    *Arrangements for the hearing*

Arrangements for the hearing are usually discussed at the preliminary meeting and should be recorded in the procedural directions. Besides confirming dates, sitting times and the venue, it is also necessary to decide which party (usually the claimant) is to reserve the necessary accommodation and to be responsible, initially, for hire and other costs incurred. In a major dispute involving counsel, witnesses of fact and experts, each party may require its own retiring room, and a retiring room may also be needed for the arbitrator in addition to the courtroom. These are all matters which are usually agreed between the parties, but may have to be determined by the arbitrator where differences arise.

The usual layout for the courtroom provides two long and parallel tables, one for each party, and a linking table across one end for the arbitrator. Witnesses, when called, sit facing the arbitrator at a separate table placed between the two parallel tables, so that their evidence may be heard and their demeanour observed by the parties and by the arbitrator. For smaller arbitrations a single long table is sometimes used, with the arbitrator sitting at one end, one party's representatives on each side of the table, and the witness giving evidence at the far end. In many cases space is needed for the large number of files and possibly drawings to be referred to while evidence is being taken: and this may require either that the tables are large enough to provide space for them, or that shelving is provided around the tables.

## 5.13    *Conduct of the interlocutory stages*

The arbitrator should at all times exercise the greatest care in ensuring that he is, and is seen to be, impartial in his dealings with both parties. He should not have any communication with either party without the knowledge of the other, so that all of his letters are always addressed to both parties, and he should ensure that letters received from a party are copied to the other party. He should have no communication by telephone with either party: urgent matters may be dealt with by fax, copies of letters to and from each party being sent to the other party. Similarly he should not meet either party except when the other party is present, and he should not accept hospitality from either party.

In dealing with applications, such as those for extensions of time, for orders requiring disclosure of documents, or for any other purpose, the arbitrator should ensure that each party has an opportunity to address him, either in writing or orally, before he reaches his decision. Even where an application for an interlocutory meeting is opposed by the other party the arbitrator should grant the application, since the only possible injustice to the opposing party – namely incurring unnecessary costs – may be remedied in an award of costs.

## 5.14  *Preliminary questions of law*

It sometimes becomes clear at the preliminary meeting or during the later interlocutory stages that the success of a claim, or of a substantial part of it, depends upon a question of law which, if determined as a preliminary issue, might save unnecessary delay and costs. Where for example a defence is based on an allegation that the claim must fail because of the operation of the Limitation Acts, such a defence, if successful, would obviate any need to proceed further.

The arbitrator has jurisdiction to determine questions of law, and he may do so either from his own knowledge or, where appropriate, he may take legal advice. Where both parties are represented by counsel, however, it is to be expected that each counsel will explain to the arbitrator the basis of his submission and the law relating to it. The arbitrator's task is then to decide which of the opposing opinions is correct. He may well take the view that there is nothing to be gained by hearing another counsel addressing the same issues, because at the end of the day the arbitrator must make his own decision: he must not delegate that decision to someone else.

A third means of determining a question of law may sometimes be available under section 45 of the 1996 Act, which provides, in limited circumstances, for determination of preliminary questions of law by the court. Leave to obtain such determination is granted only where the applicant satisfies the court first that the question of law substantially affects the rights of one or more of the parties, and second that the application is made either with the agreement of all other parties to the proceedings, or the applicant satisfies the court that determination of the question is likely to produce substantial savings in costs. The applicant must also satisfy the court that the application was made without delay. The arbitrator may continue with the proceedings, and make an award, while the application to the court is pending, unless the parties agree to the contrary.

## 5.15  Small claims

Where the claim is small – and a suggested definition of such a claim is one of less than five figures at 2000 prices – one of the principal aims of the arbitrator should be to ensure that costs are not allowed to become disproportionate to the sum in dispute. Some such claims may fall within the provisions of sections 89 to 91 of the 1996 Act, which refer to the Unfair Terms in Consumer Contracts Regulations 1994, which have since been revoked and replaced by the Unfair Terms in Consumer Contracts Regulations 1999 (SI 1999 No 2083 (see Appendix E)) in order to comply with United Kingdom obligations as a member of the European Union.

The 1999 regulations apply to 'unfair' terms in contracts between a seller or supplier and a consumer, and provide that terms are unfair if they have not been individually negotiated: i.e. if they have been drafted in advance by the seller or supplier. Furthermore the onus of proof that the terms have been individually negotiated rests upon the seller or supplier. The effect of such an 'unfair' term is that it is not binding upon the consumer, but the contract continues to bind the parties if it is capable of continuing in existence without the unfair term.

Under the Unfair Arbitration Agreements (Specified Amount) Order 1999 (SI 1999 No 2167 (see Appendix F)), which came into force on 1 January 2000, the limit to the value of claims for the purpose of section 91 of the 1996 Act, in England and Wales is increased to £5,000.

For claims not falling within these regulations, either because the monetary limit is exceeded or because the regulations are inapplicable, the objective of minimising costs is more likely to be achieved where the parties agree not to appoint any form of professional representation, legal or technical, and to the arbitrator acting inquisitorially. Section 34(2)(g) of the 1996 Act empowers the arbitrator, subject to the right of the parties to agree otherwise, to 'take the initiative in ascertaining the facts and the law'. Lay parties usually have little idea how to present their cases or to challenge the cases of their opponents, and in such circumstances the adoption of an inquisitorial procedure may help to ensure that the parties' lack of procedural knowledge does not prejudice a fair determination of the dispute.

Although the arbitrator is empowered under the 1996 Act to decide the above procedural matters in the absence of agreement by the parties to the contrary, it is suggested that he should invite the parties, at the commencement of the arbitration, to agree to those matters: namely where appropriate to waive their right to appoint professional representatives, to adopt an inquisitorial procedure, and to adopt the use of written evidence and submissions. In addition it may be necessary to arrange for inspection of real evidence, such as alleged defects in a building: and this can conveniently be combined with a meeting at which the arbitrator seeks answers to any questions he may have relating either to the written submissions or to the real evidence. The format of a letter suggesting such procedure is given in SD/14 in Appendix A.

# CHAPTER SIX
# EVIDENCE

## 6.1 Synopsis

Evidence is defined in *Osborne's Concise Law Dictionary* as meaning

'all the legal means, exclusive of mere argument, which tend to prove or disprove any matter of fact, the truth of which is submitted to judicial investigation'.

The kinds of evidence, and the rules relating to its admissibility, relevance and weight are outlined in this chapter.

## 6.2 Statutes

Section 34(1)(f) of the 1996 Act empowers the parties to agree whether to apply strict rules of evidence, or any other rules, as to the admissibility, relevance and weight to be given to material adduced in evidence, and the manner of its presentation. In the absence of any such agreement the arbitrator may determine that question. Where rules are to be applied, the Civil Evidence Act 1995 is the relevant statute, together with the Civil Evidence Acts of 1968 and 1972, to the extent that they are not repealed by the 1995 Act. A decision as to the adoption of rules should normally be made at the preliminary meeting (see Chapter 5, subsection 5.3.3).

## 6.3 Kinds of evidence

Evidence may be categorised in several different ways, depending upon its form, content and quality. The form of the evidence may be documentary, oral or real. Documentary or oral evidence may be of fact or of opinion, and it may be primary or secondary evidence. Documentary evidence is usually by far the most important in construction contract arbitrations.

## 6.3.1 Documentary evidence

Within this category are included letters, documents, drawings, photographs, computer printouts, tape recordings, and other forms of copying or of electronic storage. Documents in all of these forms must, where disclosure of documents is agreed or ordered, be listed; either individually, or where agreed or directed by the arbitrator, by classification. In general the original document – for example, the letter which actually passed between the parties – constitutes primary evidence, and should be adduced and made available for copying where necessary. Where the original document has been lost or is not available for some reason, a file copy of the document may be adduced as secondary evidence. It should however be recognised that the method by which the original document was produced is immaterial: primary evidence may be in the form of a photocopy of a letter, or of a dyeline print of a drawing, if that was the form of the document sent by one party to the other. In *Fairclough Building* v. *Vale of Belvoir Superstore* (1990) 56 BLR 74 it was held in the Official Referee's court that the drawings sent to the builder, which were copies of original drawings, were the actual drawings to which the builder had to work. Hence they were primary evidence, and it was unnecessary to prove what were the contents of the original drawings in the respective drawing offices.

## 6.3.2 Oral evidence

Witnesses of fact are usually required to give their evidence under oath, the arbitrator being empowered to administer the oath or take an affirmation under section 38(5) of the 1996 Act. A witness who knowingly makes a false statement under oath commits perjury, which is a criminal offence, and punishable by a fine, or by imprisonment, or both. However in many construction contract arbitrations the unreliability of oral evidence arises not from witnesses' intention to mislead, but more usually from their inaccurate recollections of events which occurred possibly some years before the date when their evidence is given.

Where necessary a party may, with the permission of the arbitrator or with the agreement of the other parties, obtain through the court an order of *subpoena ad testificandum*, requiring a witness to attend the hearing for the purpose of giving evidence. Such an order may be used only where the witness is in the United Kingdom, and the arbitral proceedings are in England, Wales or Northern Ireland (section 43 of the 1996 Act). Where evidence is required from a person who cannot readily be brought to the hearing, a party may apply to bring evidence on affidavit: that is, a statement sworn before a commissioner for oaths. However evidence on affidavit is of less weight than that given by a witness present at the hearing, because the latter is available to submit to cross-examination.

### 6.3.3  Real evidence

Material objects presented for examination by the arbitrator are termed *real evidence*. They may be samples brought to the hearing, such as bricks, pieces of concrete, steel sections or similar samples, or they may be immovable objects such as bridges, dams, or buildings, which can only be inspected on the site. Where such evidence is to be adduced, the arbitrator should arrange a date and time to make his inspection, in the presence of a representative of each of the parties; these representatives should be permitted to draw the arbitrator's attention to any matters they wish him to note, but not to make any representation to him.

### 6.3.4  Evidence of fact

Unless a witness is qualified as an expert (see subsection 6.3.5) his evidence must be confined to factual matters that are within his personal knowledge. He is not permitted to express opinions. It is however often the case that the facts upon which opinions may be expressed are of more value than the opinions. For example, a statement that drawings were issued late is likely to be less convincing to an arbitrator than evidence as to the actual dates of issue of the drawings, and of the dates on which the work they depict were programmed to be commenced. Similarly a statement that a brick wall is 30 mm out of plumb and contains courses varying from 72 to 77 mm in height is of more evidential value than a subjective opinion that the quality of workmanship is poor.

However, section 3(2) of the Civil Evidence Act 1972, which is unaffected by the Act of 1995, provides for the admissibility of a

'statement of opinion by [a person called as a witness in any civil proceedings] on any relevant matter on which he is not qualified to give expert evidence, if made as a way of conveying relevant facts personally perceived by him.'

### 6.3.5 Evidence of opinion

A witness called to give evidence of opinion on matters in which he is suitably qualified is termed an *expert*, and his appointment and evidence are subject to rules contained in the Civil Evidence Acts of 1972 (as amended) and 1995, and to the Civil Procedure Rules 1998. The role of the expert is to assist the arbitrator in coming to a correct decision on the issues for which he is called to give evidence: he is not in any sense an advocate for the party by whom he is called. For this reason an expert should not accept an appointment as such by a party until he is sure that his opinion will support the case of that party. Where he finds himself to be in disagreement with the party's case he should advise the party of his opinion and recommend that the point in issue be conceded. Where he is satisfied that his evidence will support the case of the party by whom he is to be called he must still ensure that his evidence is directed towards the elucidation of the truth rather than a mere presentation of his party's case. This does not imply that he should make the opposing party's case for it; but that he should be aware of any weaknesses in his party's case and should give his evidence covering all relevant matters, including such weaknesses, with honesty and sincerity.

Where the expert is to be appointed by the arbitrator, his duty to assist in elucidating the truth is unchanged, but problems of actual or suspected bias in favour of the party by whom he is called do not arise.

The 1996 Act empowers the arbitrator, under section 37 and unless otherwise agreed by the parties, to appoint experts or legal advisers 'to report to it and to the parties'. Additionally it provides that the parties shall be given 'reasonable opportunity to comment on any information, opinion or advice offered by any such person'. The Act makes no express provision however for the appointment of experts by the parties, thus implying that an expert appointed by the arbitrator is to be the norm; although it clearly provides an opportunity for party-appointed experts to challenge the evidence of the arbitrator-appointed expert. The latter type of appointment would of course defeat the objective of the Act's provision: namely to save the need for, and cost of, more than one expert.

There are occasions upon which a single expert appointed by the arbitrator would meet the parties' needs: where for example the scale of the dispute does not warrant the appointment of two experts, or where there exists a person whose expert knowledge of the subject in question is acknowledged and respected by both parties. There are however also occasions upon which experts appointed by the parties within an adversarial framework are more appropriate. For example where there are differences of informed opinion upon a particular topic, or where expert quantity surveyors appointed by each party are needed to deal with a large number of items of claim, and may be successful in agreeing some or even most of them. In this latter situation it is clear that the parties are empowered to agree, under section 37 or more generally under section 34, as to arrangements for the appointment of experts.

An expert's qualifications need not necessarily be formal, such as university degrees or membership of a learned body. In some cases relevant experience may be the prime requirement: for example the appropriate expert in respect of an issue under Clause 12 of the ICE Conditions of Contract as to whether or not adverse physical conditions or artificial obstructions were foreseeable by an experienced contractor could well be an experienced contractor.

It often happens in construction contract disputes that witnesses of fact are also expert in that they are experienced engineers, builders, or quantity surveyors, although not called as experts in those subjects. The relaxation of the rules relating to expert evidence of a person not qualified as such, referred to in subsection 6.3.4, may enable the arbitrator to give due weight to opinions expressed by such witnesses.

## 6.4  Admissibility

The prime requirement of evidence in order that it may be admitted is that it must be relevant to the matter in issue. It is a part of the arbitrator's function to determine questions of admissibility; but where the parties are represented by counsel he can usually assume that they will raise an objection to any evidence considered to be inadmissible. Where either or both parties are not so represented, the arbitrator may find it necessary to draw attention to defects in the evidence being adduced and to invite an objection from the opposing party.

The arbitrator should not refuse to hear admissible evidence. There may however be occasions when he considers that a point is being laboured unnecessarily, when he may indicate to counsel that he has taken note of the point, and invite him to proceed to his next topic.

## 6.5  Privilege

The main categories of privileged documents likely to be referred to in construction arbitrations fall into two categories: those privileged because they contain professional confidences, and those relating to offers to settle the dispute or a part of it, where marked 'without prejudice', or 'without prejudice save as to costs'. Such documents must be included in each party's list of documents where such list is required; but the documents themselves must not be disclosed to the opposing party or to the arbitrator.

The first category, legal professional privilege, covers correspondence that passes between a party and its legal or other professional advisers such as technical experts, in connection with actual or contemplated legal proceedings. Clearly advice from such an adviser as to the likelihood of success of a claim would be damaging to the party's case if disclosed to the other party. Each party must be free to communicate in confidence with its advisers, without fear that such correspondence might fall into the hands of the opposing party.

In the second category are documents containing offers to settle could similarly be damaging to the offeror's case if disclosed to the arbitrator, since they could be construed as an admission of liability to the extent of the offer made. For that reason offers are usually marked 'without prejudice', indicating that they may not be adduced in evidence at any stage of the proceedings, or 'without prejudice save as to costs', indicating that they may be so adduced only after the substantial issues have been determined and the arbitrator has reached the stage of dealing with costs.

## 6.6  Proofs of evidence

A witness of fact is not permitted to read from a prepared statement, or even to refer to such a statement, while giving evidence. He is permitted to refer only to contemporaneous notes of the events referred to, for example his diary of those events. If he is observed to be reading from some document the opposing counsel will usually require to inspect that document in order to satisfy himself that it complies with the above requirements.

It is now standard practice for witnesses of fact to be required to prepare proofs of their evidence, and for such proofs to be exchanged between the parties before the hearing. Such proofs are often deemed to stand as evidence-in-chief of the witness, and the witness, after taking the oath, will be asked by his party's counsel to identify his statement, to confirm that it is true, and to state if he wishes to add to or to withdraw anything in it. Unless he finds it necessary to make any such modifications – and he may need to do so if, for example, some previous evidence calls for his comment or reminds him of some fact that he inadvertently omitted – he is then available for cross-examination. The advantages of this procedure are first that it saves the time that would otherwise be spent in examination-in-chief; second that it assists the arbitrator by providing a written record of the witness's evidence-in-chief; and third that it assists the opposing counsel in preparing his cross-examination. Where the proof of evidence of an opposing party's witness differs from a party's own version of the facts, the reasons for the discrepancy may be investigated in advance of the hearing, and where necessary the party's version of the facts may be amended if it appears that the proof of evidence of the opposing party's witness has brought to light errors in a witness's proof of evidence. An example of a proof of evidence is given in SD/15 in Appendix A.

## 6.7  Calling witnesses

In general it is a function of the parties to call witnesses, both of fact and of opinion, and the parties' intentions will usually have been discussed and agreed, or determined by the arbitrator, at the preliminary meeting or at a later stage, after statements of case have been exchanged and the issues have been identified. In some cases it may be necessary for the parties to revise their intentions in such matters in the light of the information gained during the exchange of statements of case and disclosure of documents. Provision is also made under section 34(2)(g) of the 1996 Act for the arbitrator to take the initiative in ascertaining the facts and the law, and where the parties agree to this course, or the arbitrator so directs, it is open to him to call witnesses. In most cases however it is sufficient, where the parties are legally represented, for the arbitrator to draw their attention to any apparent omissions from the parties' lists of witnesses. For example, an enquiry: 'shall we be hearing evidence from the engineer?' is usually sufficient to alert the employer's counsel to the need for such evidence.

## 6.8  Burden and standard of proof

The burden of proving an assertion lies upon the party by whom it is made, except where the other party admits its truth. Having provided such proof, the burden shifts to the party seeking to disprove the assertion, if it seeks to do so, or alternatively to show that the point is not material to the main issues in the arbitration.

The standard of proof required in arbitration is that applicable to civil actions in the courts: namely on the balance of probability. It is for the claimant to satisfy the arbitrator that its allegations are more probably true than those of the respondent in order to succeed in its claims. Conversely it is for the respondent to satisfy the arbitrator that its allegations are more probably true than those of the claimant in order to succeed in its defence. The standard of proof required in civil actions and in arbitrations may be contrasted with that required in the criminal courts, in which the burden on the prosecution is to prove guilt to a much higher standard; namely 'beyond reasonable doubt' in order to obtain a conviction.

# CHAPTER SEVEN
# THE HEARING

## 7.1 Synopsis

Where a hearing is included in the arbitration procedure it is open only to participants in the arbitration. The procedure and sequence of events usually follows a standard pattern, which may however be varied to suit a particular reference.

## 7.2 Attendance

Notwithstanding that the date, time and venue of the hearing have usually been agreed, the arbitrator should issue a formal notice giving that information and requiring both parties to acknowledge its receipt. Where it appears likely that a party may fail to appear – for example where the party has failed to acknowledge any of the arbitrator's earlier letters – he should add a warning such as:

'...and I give formal notice that, in the event of failure by the Respondent to appear, either in person or by representation, I shall, on the application of the Claimant, proceed *ex parte* to hear the claim pursuant to my power under section 41 of the Arbitration Act 1996, and thereafter to make my award'.

If a party unexpectedly fails to appear at the hearing, and hence has not been given a warning of the above type, the arbitrator should wait for a reasonable period in case the party has been delayed on the way to the hearing, and should if possible check by telephone whether or not the party intended to be present. While waiting the arbitrator should of course have no communication with the party in attendance, and should preferably wait in a separate room. If and when it becomes clear that nothing is to be gained by further waiting the arbitrator should adjourn the hearing to a later date. When notifying both parties of the date and time of the adjourned hearing the arbitrator should mark his letter 'Peremptory' and should add a warning of the above type in order that he may proceed *ex parte* in the event of a further default. Costs thrown away by default of a party should generally be awarded against that party unless there is some special reason for making some other award.

In arbitration the hearing is private: attendance is limited to the arbitrator, the parties and their representatives and witnesses, both of fact and of opinion; and where a transcript is to be made there will also of course be a stenographer. Others may attend only with the consent of both parties and of the arbitrator, and such consent is usually given where the arbitrator or counsel seeks leave to invite a pupil wishing to gain experience of arbitration procedure, on the condition that the confidentiality of the proceedings will be respected.

Where the credibility of one or more of the witnesses of fact is in question, a party may ask the arbitrator to rule that such witnesses be excluded from the hearing until the time comes for them to give their evidence. If requested to make such a ruling the arbitrator should of course invite the opposing party to comment on the matter before making his decision. Such a ruling may be especially appropriate where it is believed that a party's witnesses of fact may have discussed and formulated a party line on some issue: for example the cause of a delay to the construction work.

## 7.3  Courtesy

It has in the past been the practice of some arbitrators to ensure that the parties and their representatives are all in attendance and seated before they enter the courtroom; and on the entrance of the arbitrator, for all of those present to rise, as is customary in court procedure. More recently the practice is not generally used, possibly because of a desire to distinguish arbitration from litigation. A less formal approach is often preferred both by the arbitrator and by the parties; for example in that counsel and witnesses usually remain seated while evidence is being taken, although some counsel prefer to stand.

The arbitrator should be addressed as 'Sir' and should be referred to in the third person as 'the arbitrator': not as 'Mr Smith'.

## 7.4  Challenges to arbitrator's jurisdiction

A practice was sometimes adopted before the enactment of the 1996 Act by a party seeking to delay or disrupt the proceedings, of challenging the arbitrator's jurisdiction at the opening of the hearing. By so doing the party was sometimes able to obtain an adjournment, delaying the commencement of the hearing by weeks or by months. Such tactics are unlikely to succeed under the provisions of section 73 of the 1996 Act, which provides *inter alia* that a party who takes part or continues to take part in arbitration proceedings without making, within such time as is allowed, any objection to the proceedings, may not raise that objection later, before the arbitrator, unless he shows that he did not know, or could not have discovered at the appropriate time, the grounds for such objection.

## 7.5 Representation

The parties should have indicated their intentions as to their representation at the preliminary meeting (see Chapter 5, sub-section 5.3.3) or, where no such meeting is held, in correspondence, and their intentions should have been recorded in the arbitrator's Procedural Directions. Should a party appear at the hearing with some different form of representation from that notified earlier – for example, by leading counsel where the stated intention was to appoint junior counsel – the arbitrator should invite the other party to apply for an adjournment in order that it can arrange similar representation. In such a case the costs resulting from the adjournment should be awarded against the party who failed to give notice. There are, however, several other ways in which such a difficulty may be resolved; e.g. in the example quoted above by the defaulting party dispensing with the services of the counsel to whom objection is taken and continuing in the hearing with junior counsel; or by the other party waiving its objection to its opponent's representation. In most cases however, counsel will be well aware of, and will respect, the procedural rules, so that the likelihood of the arbitrator having to deal with such a situation affecting counsel is remote. The same rule does however apply generally: for example to the appointment of experts where no notice has been given.

## 7.6 Procedure

### 7.6.1 Sequence: calling witnesses

The arbitrator should open the hearing by outlining the events that have led to it: the identity of the parties and of the contract between them; the arbitration agreement therein; the occurrence of a dispute; the manner in which he was appointed to be arbitrator and his acceptance of that appointment. He may also ask each party to identify by name and by appointment each of its representatives, experts and witnesses of fact, or may ask each party's counsel to provide a list of such persons. It may sometimes be necessary to check at this stage that there are no unauthorised persons in the courtroom. The arbitrator then declares the hearing open, and invites the claimant's counsel to open his case.

The claimant's counsel introduces himself and his 'learned friend' who will represent the respondent. Where, as is now usual, there has been an exchange of written opening submissions, counsel may need only to identify his submission, check that the arbitrator has had time to read it, and proceed to call his witnesses. In many cases however, and especially those of a complex nature, counsel may wish to emphasise the parts of his opening which he considers to be especially important, and possibly to comment on his opponent's opening where such a document has been issued in advance of the hearing. Where the parties have agreed, or in the absence of agreement the arbitrator has imposed – usually at the preliminary meeting – a limitation on the time available to each party for the presentation of its case, the use of the available time is at the party's discretion. In other cases, and especially where hearing time has not been limited, the arbitrator may seek to discourage a long opening speech by counsel, on the basis that the written submission is intended to obviate the need for such oral presentation.

Following upon the claimant's counsel's opening, counsel calls each of his witnesses in succession. Each witness, when called, proceeds to the witness chair, and takes the oath or affirms. The arbitrator should have prepared himself by having available a Bible, an Old Testament, and a card on which is typed the words of the oath on one side and of the affirmation on the other: the forms of which are:

'I swear by Almighty God that the evidence I shall give touching the matters in dispute in this reference shall be the truth, the whole truth, and nothing but the truth.'

'I solemnly, sincerely and truly affirm and declare that I will true answers make to all such questions as shall be asked of me touching the matters in difference in this reference.'

Witnesses who are Christians should hold the Bible in their right hand while taking the oath, while those of the Jewish faith should repeat the same words, holding the Old Testament. Others, and those who so prefer, should be permitted to recite the words of the affirmation.

Each witness's evidence begins with examination-in-chief, and is followed by cross-examination, and then by re-examination. When all of the claimant's witnesses have given their evidence the respondent's counsel opens his case and then calls each of his witnesses in turn for examination-in-chief, cross-examination and re-examination, as in the case of the claimant's witnesses. When all of the respondent's witnesses have given their evidence the respondent's counsel makes his closing address, and then the claimant's counsel makes his closing address.

### 7.6.2   Proofs of evidence

Where a witness has submitted a full proof of his evidence the need for examination-in-chief should be eliminated or at least reduced to a few short statements. In such cases counsel need only ask the witness to identify his written statement, to confirm that it is true, and to state if there are any matters that he needs to add to it or to amend. It may in some cases be necessary for the witness to comment on some matter that has come to light in earlier evidence, but he should not have to elaborate upon the facts recorded in his statement if it has been properly prepared. The purpose of proofs of evidence is to eliminate the need for lengthy examination-in-chief: and counsel should be encouraged to respect that intention.

### 7.6.3   Leading questions

During examination-in-chief and re-examination counsel is not permitted to lead his witness. A leading question is one which implies the answer sought by the questioner, such as: 'was the delay caused by variations to the work?' The correct form of such a question is: 'what caused the delay?' An exception is usually permitted where the questions are not in contention and where counsel may obtain the answers he requires more readily by asking, for example:

'You are Joe Bloggs; you live at 25 High Street, Blanktown, and you were employed on the contract as site foreman from January until October 1997?'

than by asking each of the questions that would otherwise be needed.

Where counsel deviates from the rule by asking one of his own witnesses leading questions it is open to the opposing counsel to object, and where appropriate the arbitrator should allow the objection by asking counsel to re-phrase the question. That does not however deal adequately with the objection, because the witness is already aware of the answer sought by his party's counsel, and may simply give that answer. If counsel persists in asking leading questions of his own witnesses a more effective remedy may be for the arbitrator to warn counsel that he will, when weighing the evidence, take note of any leading by counsel.

A different rule applies during cross-examination, in that counsel is free to ask leading questions of an opposing party's witness, and indeed needs to do so. This is because one of the tasks of cross-examining counsel is to put to each witness having knowledge of a particular allegation the opposing version of the facts, so that the witness may comment on that version. Hence, for example, counsel may ask during cross-examination:

'I put it to you that the delay to which you refer was not caused by variations, but by the fact that there was insufficient plant and labour on the site?'

A question in that form invites the witness to express his own knowledge of the matter in question.

Re-examination provides an opportunity for counsel to question his own witnesses on matters raised during cross-examination, usually with the objective of correcting any erroneous impressions that the witness may have given to the opposing counsel. Cross-examination may become a very trying experience, especially where the witness's recollection of events is not as reliable as he might seek to imply, and it is not unusual for a witness to become flustered when subjected to even mildly hostile questioning. His own party's counsel will seek during re-examination to remedy any damage caused to the witness's credibility by his opponent. He may not raise fresh matters, because by so doing he would deny opposing counsel's right to test those matters, but he may be able, by skilful questioning, to correct misleading impressions.

### 7.6.4  Arbitrator's questions

At the conclusion of re-examination counsel usually asks the arbitrator if he has any questions for the witness before he is released. The arbitrator is of course free to ask questions at any time while evidence is being taken, and he should do so whenever necessary: for example to clarify a witness's meaning or to seek an explanation of an apparent inconsistency. He should however try to avoid interrupting counsel's flow of questions, especially where those questions are leading towards a conclusion that may not initially be clear either to the witness or to others. It may however save time if the arbitrator asks his questions on a particular topic while it is being explored by counsel, especially where it is necessary for documents to be referred to during the questioning. The practice of remaining a 'po-faced arbitrator', debated some years ago, has been firmly rejected: while on the other hand attempts by some arbitrators to take over the functions of counsel are equally objectionable. In many cases questions which the arbitrator may feel inclined to ask, and especially those covering new topics, may already be in counsel's list of questions to be asked, and it is far better, in adversarial proceedings, that counsel should be free to ask those questions in his own way.

Different rules apply where the parties have agreed, or the arbitrator has decided, to adopt an inquisitorial procedure, pursuant to section 34(2)(g) of the 1996 Act. In such a case the arbitrator performs all of the tasks that would otherwise fall upon the parties' advocates: namely calling witnesses and conducting examination-in-chief, cross-examination and re-examination. His objective in questioning witnesses is of course to ascertain the truth and for that reason he should not ask leading questions. He should however give each witness an opportunity to comment upon the opposing party's version of the facts, and in order that he may do so proofs of evidence of all of the witnesses to be called are likely to be helpful, if not essential, to the arbitrator. In addition the arbitrator should, before the hearing, study carefully both parties' submissions and the proofs of evidence of all witnesses, in order that he may identify the issues to be resolved and the versions of facts to be put to each witness where there is a discrepancy between his evidence and that of another witness.

Where the arbitrator appointed as being a person having technical knowledge relevant to the matters in dispute finds that the evidence being given by a witness, and especially an expert, is inconsistent with his own knowledge and/or experience, he must draw the witness's attention to the inconsistency, and invite him to comment on it. Should he fail to do so he will in due course find himself in the unenviable position of having to make an award which is either, in the light of his own knowledge, incorrect, or which is unfair, taking account of evidence taken from himself and not exposed to challenge at the hearing. In *Fox and Others* v. *P.G. Wellfair* (1981) 19 BLR 52 (see Chapter 8, section 8.2) the arbitrator used his technical knowledge in coming to a different evaluation of a claim from that given in evidence by the claimant's experts. In so doing the arbitrator had effectively taken evidence from himself: but he had failed to expose that evidence to challenge by the party present at the hearing. His failure to do so was held in the Court of Appeal to constitute misconduct of the reference (now termed 'serious irregularity'), in respect of which the arbitrator's award was set aside.

# CHAPTER EIGHT
# THE AWARD

## 8.1  Synopsis

In making his award the arbitrator's primary objective is to define clearly, unambiguously, justly and enforceably the actions to be taken by one or both parties in order to resolve the matters in dispute. His secondary objective is to satisfy the parties, and in particular the losing party, that the award is just. This chapter covers the procedure for making an award, types and format of awards, the manner in which the availability of the award is notified to the parties, and the arbitrator's powers to correct errors and to make additional awards.

## 8.2  Procedure

The award must be based on evidence *adduced at the hearing*; or, where the parties have agreed to waive their right to a hearing, on the documents submitted to the arbitrator in accordance with the agreed procedure. Where the procedure includes a hearing the information upon which the arbitrator must base his award comprises the following.

(1)  Documentary evidence in agreed bundles, some or all of which may have been referred to at the hearing.
(2)  Oral evidence adduced from witnesses of fact and recorded in their proofs of evidence, in the arbitrator's notes and retained in the arbitrator's memory; possibly supplemented by a transcript or a tape recording of the hearing.
(3)  Written reports and oral evidence of expert witnesses.
(4)  Real evidence, if any, inspected before, during or after the hearing.
(5)  Counsel's submissions, as contained in opening and closing addresses, and as expressed orally at the hearing.

Another source of information, which the arbitrator should use with the greatest of care, is his own knowledge and experience, constituting evidence which the arbitrator effectively takes from himself. Where such evidence is in conflict with evidence adduced by either or both of the parties, the arbitrator *must* draw the parties' attention to it, and to the conflict he finds, while at the hearing. Should he fail to do so, the arbitrator will find himself to be faced with the options of either (1) making an award which in the light of his own knowledge is incorrect, or (2) making an award which he believes to be correct, but which is unjust because it takes account of evidence not presented for challenge at the hearing.

In *Fox and Others* v. *P.G. Wellfair* (1981) 19 BLR 52 the Court of Appeal, under the late Lord Denning, then Master of the Rolls, unanimously dismissed an appeal against a High Court order removing the arbitrator and setting aside his award on the ground of misconduct. The arbitrator, who was an architect and a barrister, had used his special knowledge in reaching a conclusion which was in conflict with unchallenged evidence adduced by the claimant, without providing an opportunity to challenge that knowledge. In an *ex parte* hearing at which only the claimants' representatives were present, the arbitrator had listened carefully to the evidence of two witnesses of fact and four experts, and had intervened where he required clarification. He had not, however, felt it to be a part of his duty to indicate at the hearing that he did or did not accept any particular evidence. On his understanding of his duty the arbitrator had rejected the evidence of four experts, all eminent in their own specialisations, indicating that the cost of remedial works to a block of flats in which defects had given rise to the claims was £93,000. Instead he had awarded £13,000 in respect of the claims. He had done so without giving any indication to the witnesses or to counsel for the claimants that he was rejecting the claimants' evidence. If he had given them any indication of his views, they would have been able to correct them.

## 8.2.1 Issues of fact

It is a part of the arbitrator's duty to determine questions both of fact and of law. Fact in this context includes matters of opinion: for example it includes the question that frequently arises from contracts under the ICE Conditions of Contract of whether or not the physical conditions or artificial obstructions encountered by the contractor were such as might reasonably have been foreseen by an experienced contractor. Where, as is often the case, the facts are in dispute, the arbitrator must weigh the evidence of each witness dealing with the issue. In doing so, he is usually assisted by skilful questioning of witnesses by counsel, especially during cross-examination, when weaknesses in recollection or in opinion may be exposed.

## 8.2.2   Issues of law

Having determined the facts, the arbitrator's next step is determine issues of law; and here again counsel's submissions are usually helpful, especially to a legally lay arbitrator, in coming to a logical and legally correct decision. Counsel will usually explain to a lay arbitrator the significance of questions of law on which they are agreed, and make their individual submissions on matters that are in contention.

Section 46(1) of the 1996 Act introduces a provision under which the dispute is to be decided either

'(a) in accordance with the law chosen by the parties as applicable to the substance of the dispute; or
(b) if the parties so agree, in accordance with such other considerations as are agreed by them or determined by the tribunal.'

The first of these alternatives gives the parties freedom to agree as to the application of a system of law: for example the substantive laws of a defined country. The latter provision is most unusual in current English law; although in the past a concept of equity (fairness or natural justice) was kept distinct from the basic common law and applied in different courts. It is conceivable that the parties to a construction contract dispute might agree that questions of law be determined in accordance with, for example, 'usual practice in the construction industry', but the adoption of such a formula would give rise to enormous problems of definition. Moreover construction contract law is well reported through such publications as the Building Law Reports and further expounded in such standard texts as *Hudson* or *Keating*, relied upon and referred to by judges and arbitrators alike. It is difficult to envisage any practical advantage that would be gained from the adoption of any basis of determination of issues of principle that may arise other than the application of the law chosen by the parties. It is submitted that the arbitrator should, unless there is some special reason to the contrary, advise the parties against the adoption of section 46(1)(b) of the 1996 Act.

Where issues arise on questions of law, usually from the construction either of clauses in a standard form of contract or from some *ad hoc* variation or addition to the standard form, each counsel will draw the arbitrator's attention to any relevant case law there may be on the issues that arise, and on which he relies. Having heard both counsel's submissions, the options open to the arbitrator for determining the issue are as follows.

(1) Relying upon his own understanding of the issue and of counsel's submissions.
(2) Taking legal advice.
(3) Application to the court for determination of the point of law pursuant to section 45 of the 1996 Act.

In many cases the arbitrator may prefer to adopt the first of the three options. Where the law has been explained by learned counsel before him, the arbitrator may take the view that hearing another counsel, however learned, will add little to his understanding of the issue. Furthermore it is he, the arbitrator, who must decide the issue; he must not delegate that decision to the legal adviser.

Where the arbitrator decides to adopt the second option it is advisable that he invite counsel before him to define the questions of law to be determined and to nominate, preferably by agreement, the counsel from whom advice should be sought. Failing such agreement the arbitrator must make his own choice, but in doing so he should take note of the nominees proposed by the parties.

The third option is available only where application to the court is made with the agreement of all of the parties to the proceedings; or where it is made with the permission of the arbitrator, and the court is satisfied that determination of the question is likely to produce substantial savings in cost, and that the application was made without delay. In most cases these restrictions necessarily imply that the questions of law are identified and dealt with before the hearing: otherwise little saving will result from determination of the questions at that stage of the proceedings, instead of challenging the award. Where this course is adopted the arbitrator should invite the parties to define the questions of law to be put to the court for determination. In such cases the arbitrator may continue with the proceedings and make an award while the application to the court is pending, unless the parties agree otherwise.

Finally, having determined both the facts and the law to be applied to those facts, the arbitrator must decide as to the validity of the claims and of the counterclaims if any, on the remedies to which the parties are entitled, and on his award of interest and costs.

## 8.3 Types of award

Section 48 of the 1996 Act defines the several powers exercisable by the arbitrator in the absence of agreement of the parties to the contrary as regards remedies.

(1) To make a declaration on any matter to be determined.
(2) To order the payment of a sum of money, in any currency.
(3) To order a party to do or to refrain from doing anything.
(4) To order specific performance of a contract '(other than a contract relating to land)'.
(5) To order the rectification, setting aside or cancellation of a deed or other document.

In addition, section 51 provides for an agreed award where the parties settle their dispute.

### 8.3.1 Declaratory awards

Awards of this first type are often made where, for example, an issue arises as to liability, and it is decided that it should be determined separately from issues of quantum. By so doing the need to spend time and money in determining quantum issues will arise only if liability is established.

### 8.3.2 Monetary awards

Awards of the second type, termed monetary awards, are the most common, and usually require the determination of issues both of liability and of quantum, in respect of claims and possibly counterclaims, unless liability has already been dealt with in a declaratory award.

### 8.3.3 Injunction awards

The need for an award of the third type rarely arises. In many cases the arbitrator may consider it more appropriate to give his directions in the form of an order rather than an award, and his powers for giving such orders are available under sections 38, 39, 40 and 41 of the 1996 Act.

### 8.3.4 Performance awards

An award of specific performance is sometimes requested by the parties; but in general, unless there are special reasons why the original contractor for the work should be given the task of remedying defects, a monetary award is to be preferred. Cases do however sometimes arise where special techniques or equipment used by the original contractor are not available elsewhere, and uniformity of finish can be achieved only by employing that original contractor for the remedial works. Again, it may in some cases be thought by a building owner that the divided responsibility that could result from employing another contractor to carry out remedial works should be avoided.

Where both parties request a performance award the arbitrator has no option but to accede to their wishes. He should however take several precautions against problems that may arise. First, the work to be done should be specified by agreement of the parties: it is no part of the arbitrator's function to act as the engineer or the architect in defining the remedial work. Second, the date of commencement of the specified work, and the time for its completion (subject to usual provisions for extension of time) should be specified, again preferably by agreement. Third, arrangements must be made for access to perform the work, and it may be necessary to define arrangements and responsibility for security of buildings etc. while the remedial works are carried out. Finally, provision should be made for the determination, by the arbitrator, of any dispute that may arise as to adequacy of performance: because a performance award often gives rise to a further dispute as to the quality or extent of the work carried out. It follows that the arbitrator should be careful to ensure that his performance award is not, either expressly or impliedly, a final award. A title such as 'Award as to performance of remedial work' may be considered suitable.

### 8.3.5 Rectification awards

The fifth type of award, namely that of rectification, is an innovation in the 1996 Act, although it was, under earlier legislation, sometimes available to an arbitrator where the contract so provided. Where a party alleges that a written contract does not truly reflect the parties' agreement, it is open to the party to seek an award of rectification in order to correct the mistake or inaccuracy. The principles by which such an issue is determined are similar to those relating to more usual types of dispute. Evidence may be adduced by the parties to indicate the background of events leading to the contract; and counsel may make submissions as to statute and case law relevant to the disputed matters.

### 8.3.6 Agreed awards

Section 51 of the 1996 Act provides rules, unless otherwise agreed by the parties, as to procedure where the parties settle their dispute. If the parties so wish and the arbitrator does not object, the settlement shall be recorded in the form of an agreed award. Such an award has the same status and effect as any other award, and the general provisions of sections 52 to 58 apply to an agreed award. Unless the parties have also settled the award of costs, the provisions of sections 59 to 65 of the 1996 Act continue to apply.

In general the parties should be urged to require that their settlement be incorporated in an agreed award. The benefits of so doing are first that it defines the matters that have been settled, obviating any risk that they may be resurrected at some later date. Second, an agreed award may be enforced in the same manner as any other award, should the paying party default in its obligations. Finally it terminates the reference, removing any uncertainty as to whether or not the arbitrator remains seized of the reference. (See for example, SD/16 in Appendix A.)

## 8.4 Format of the award

It has become general practice, since the 1979 Act became law, for arbitrators to include reasons either within their awards or as appendices thereto. Under section 52(4) of the 1996 Act the inclusion of reasons is mandatory except where the parties have agreed to dispense with reasons. Where the parties have so agreed, their agreement is deemed, under section 69(1) of the 1996 Act, to constitute an exclusion agreement (see Chapter 2, subsection 2.7.4).

The format of an arbitration award may vary widely as between one dispute and another, and as between one arbitrator and another. While it is possible to define what may often be included in an award, it should not be thought that the usual format is necessarily correct in every case; or that other formats are necessarily incorrect. With these provisos the usual contents of an award (see, for example, SD/17 in Appendix A) are

- headings;
- recitals;
- the contract: form etc. and recital of provisions relevant to disputed issues;
- chronology of events giving rise to the dispute;
- issues in dispute and the arbitrator's findings of fact and of law;
- the claims and the arbitrator's findings in respect of each claim;
- the counterclaims and the arbitrator's findings in respect of each;
- interest on sums awarded;
- directions as to payment or performance, or declaration as to parties' rights;
- directions as to costs (unless dealt with in a separate award);
- signature, seat of arbitration, and date.

### 8.4.1 Headings

Awards should carry the standard introduction 'In the matter of the Arbitration Act 1996 and in the matter of an arbitration between ... followed by identification of the claimant and the respondent. Although such identification is sufficient to enable the arbitrator to refer simply to 'the claimant' and 'the respondent' within the body of the award, the arbitrator may prefer to use the name or an acronym of each party, after having defined the terms he intends to use, in the interests of brevity.

Finally, there is a heading 'Award', or, where appropriate, the type of award. Unless the award is described otherwise it is deemed, under section 58(1) of the 1996 Act, to be final. It does not however necessarily terminate the arbitrator's jurisdiction immediately, because section 57(3) of the 1996 Act provides that the arbitrator may, either on his own initiative or on the application of a party, within 28 days of the date of the award or such longer period as the parties may agree:

'(a)  correct an award so as to remove any clerical mistake or error arising from an accidental slip or omission or clarify or remove any ambiguity in the award, or

101

(b)   make an additional award in respect of any claim (including a claim for interest or costs) which was presented to the tribunal but was not dealt with in the award.'

Section 14 of the 1950 Act made provision for interim awards, which dealt with part of the matters in dispute, but not all of them. Under the 1996 Act the term 'interim' is eschewed in the belief that it is confusing, and arguably a misnomer. The term was intended, under the earlier legislation, to indicate that it left some matters to be determined later, and hence that it did not terminate the arbitrator's jurisdiction. It did however finally determine all such matters as were included in the award, and the confusion is believed to relate to a belief by uninformed persons that the matters determined in an interim award were subject to review in a later award.

There remains, however, a very real need, especially in major disputes arising from construction contracts, for provision for awards covering some but not all of the matters in dispute, and for a term to indicate that the ambit of the award is so limited. Various commentators have suggested the use of the word 'partial', which it is submitted would be even more susceptible to misunderstanding than 'interim', in that it would imply that the award lacked impartiality. Section 47 of the 1996 Act provides *inter alia* for the making of 'more than one award at different times on different aspects of the matters to be determined'. Where the arbitrator does so he must specify in his award the issue, or claim, or part of a claim which is the subject matter of the award: but the Act provides no guidance as to the title of such awards.

Section 58(1) of the 1996 Act provides that

'Unless otherwise agreed by the parties, an award made by the tribunal pursuant to an arbitration agreement is final and binding both on the parties and on any persons claiming through or under them.'

This section emphasises the need for some qualification to the title 'Award' in cases where some of the matters in dispute remain to be determined. It is suggested that suitable headings might include:

- award save as to costs;
- award as to liability;
- award as to quantum;
- award in respect of claims relating to delay;
- award in respect of claims relating to variations.

Where awards have previously been made covering part but not all of the matters in dispute, and have been entitled as above or in some such terminology, it is suggested that the term 'final award' (see, for example, SD/18 in Appendix A) should be used when all other matters have been determined. Where all of the matters in dispute are dealt with in a single award, the term 'Award' is sufficient, because such an award is deemed under section 58(1) to be final.

### 8.4.2   Recitals

The purpose of recitals is to outline the various steps that led to the arbitrator having the jurisdiction and the information from which he is able to make the award. Recitals usually commence with the word 'Whereas:' under which are listed items of the following type.

(1) Identification of the contract from which the dispute arose and its purpose.
(2) The existence within the contract of an arbitration agreement; the provision therein for appointing the arbitrator, and of any procedural rules.
(3) The occurrence of a dispute.
(4) The manner in which the arbitrator was appointed.
(5) The arbitrator's acceptance of the appointment.
(6) The manner in which the arbitrator's terms were agreed.
(7) A brief outline of the interlocutory proceedings and of any procedural rules adopted after the occurrence of the dispute.
(8) The hearing: its location, date of commencement and duration.
(9) The parties' representation at the hearing: the names of counsel and of their instructing solicitors.

It may sometimes be necessary to refer also to any unusual occurrences or features of the interlocutory proceedings, where they affect the award, (for example failure of a party to attend a meeting or hearing or to comply with the arbitrator's directions), the need to inspect real evidence on site, any special agreement between the parties as to the award, or agreement as to limitation of costs.

### 8.4.3   The contract

The constitution and date of the contract will in many cases already have been outlined in the recitals (see subsection 8.4.2, point (1) above). Where necessary, further details should be provided. In addition it is usually necessary or at least desirable that the form of the contract, the edition and date of any standard form of contract incorporated, and details of any modifications to the standard form, should be stated. Thereafter it is usually desirable to set out the text of the particular clauses of the contract that are relevant to the matters in dispute.

### 8.4.4 Chronology of events

A narrative of the events leading to the dispute, although arguably unnecessary to the parties, who are already well aware of those events, should nevertheless be provided. Not only does it provide confirmation that the arbitrator is aware of the relevant facts, it also provides background information to others who may be called upon to deal with any further actions that may arise from the award, such as judges hearing applications to the High Court for leave to appeal.

In large and complex disputes it is often helpful to divide the chronology into stages: for example the pre-tender period, negotiations after submission of the tender, the commencement of work and the different phases of the construction work, and the defects correction period.

### 8.4.5 Issues in dispute

Following upon the exchange of statements of case the parties' advocates should have identified the issues that arise between the parties, and have listed them in order that the arbitrator may know the matters he is being asked to determine. Such issues may be either of fact or of law; but more commonly the latter. The need for identification of issues of law has been emphasised by the appeal procedure contained within section 69(3) of the 1996 Act, under which leave to appeal on a question of law is given only if, *inter alia*, the court is satisfied that 'the question is one which the tribunal was asked to determine.' Hence it is important, if an appeal is likely to arise, that the issue should have been clearly identified by the parties' counsel, preferably before the hearing is commenced.

### 8.4.6 Claims and counterclaims

Where, as is often the case in major construction contract disputes, there are a large number both of claims, and possibly of counterclaims, to be determined, the facts and the law relating to each should have been clearly set out by the claimant's (or, in the case of counterclaims, the respondent's) counsel. Evidence in support of each item should have been identified and where necessary summarised in counsel's opening and closing addresses.

The arbitrator's task at this stage depends upon the number and complexity of the claims before him, and the extent to which the issues affecting each claim have been separately identified and determined; for example under subsection 8.4.5. In general, the subject of each claim and the relevant facts found by the arbitrator should be stated. In addition the arbitrator's holdings on relevant issues of law should be stated, if not dealt with earlier in the award. Similar principles apply to the manner in which counterclaims should be dealt with in the award as those applicable to claims.

### 8.4.7 Interest on sums awarded

Where, as is often the case, the arbitrator finds that sums of money awarded to either party are sums which should have been paid possibly months or years earlier, he may, and in general should, award interest from the date on which each sum awarded ought to have been paid, up to and beyond the date of the award. Such interest may be awarded either as contractual interest, where the contract so provides, or as statutory interest, pursuant to section 49 of the 1996 Act.

Section 49 of the 1996 Act reproduces section 19A of the 1950 Act (as inserted under section 15(6) of the Administration of Justice Act 1982) but with two important changes. First, the restriction of the arbitrator's powers to the award of simple interest has been removed by the express statement that the tribunal may award 'simple or compound interest from such dates, at such rates and with such rests as it considers meets the justice of the case'. Second, the arbitrator is now empowered to award simple or compound interest from the date of the award (or any later date) until the date of payment. The provisions of section 49 are subject to the right of the parties to agree otherwise; but it appears to be unlikely that a party expecting to recover substantial sums of money through its claims would think it either necessary or desirable to modify the sensible and just provisions of the section.

Where provision is made in the contract for the payment of interest, then in general such provision should be implemented in preference to the statutory power, as being the agreement reached between the parties. Furthermore, where the contract expressly indicates that the provisions of section 49 (or of section 19A of the 1950 Act) do not apply, then the arbitrator should give effect to the parties' agreement. Such intention must however be clear: in Mustill & Boyd's *Commercial Arbitration*, second edition, the learned authors state:

> 'The statutory power may be excluded by express agreement, but clear words are necessary to achieve this result. Probably nothing short of an express reference to the statutory power will suffice: anything less will be assumed to exclude only a right to interest under the contract.'

Where for some reason the contractual provision dealing with the award of interest is found to be inapplicable or ineffective, the arbitrator should decide whether it was the true intention of the parties that there should be no interest payable on overdue principal sums. In the unlikely event of the arbitrator being so satisfied then arguably he should not frustrate the parties' intentions by exercising his discretion under section 49 in favour of the winning party. However, the quotation above from Mustill & Boyd suggests that even an express statement in the contract that contractual interest is not payable may not be sufficient to extinguish the arbitrator's statutory power under section 49, unless that power is expressly excluded.

Where the arbitrator exercises his power under section 49 it is suggested that he should calculate interest in accordance with the provisions of the contract so far as they are relevant. Under the ICE Conditions, seventh edition, provision is made for interest to be payable on overdue sums at 2% above base rate, compounded at monthly intervals. No corresponding provision is made in the JCT form of building contract (see Appendix G).

Interest is usually calculated up to the date of the award. Thereafter a daily amount of interest may be determined to cover the period from the date of the award until the date of payment, thereby giving the paying party an incentive to prompt payment.

## 8.4.8 Directions as to payment or other remedy

In an award directing the payment of money by one party to the other the award should give clear and unambiguous directions as to the amount of payment to be made. Thus for example a direction as to payment of interest should state the sum to be paid rather than 'interest at the rate of $x\%$ for the period of . . .' which could give rise to further contention.

### 8.4.9 Directions as to costs

In many cases it is desirable that costs should be dealt with separately, for reasons given in Chapter 9. Where directions as to costs are included in the award on the substantive issues such directions should cover both the arbitrator's decision in principle as to which party is responsible for the payment of costs, and the provisions for determination of the amount of the parties' costs in the absence of agreement. The arbitrator should determine his own costs in accordance with his agreement with the parties where applicable, otherwise on a reasonable basis pursuant to section 64 of the 1996 Act.

### 8.4.10 Determination of recoverable costs

Before the 1996 Act became law it was usual for the parties' costs to be 'taxed' (i.e. 'determined' in current terminology) by a taxing master of the court; especially in cases where it was, or might be considered to be, appropriate for the parties to have appointed counsel. The arbitrator in such cases often marked his award 'fit for counsel' in order to indicate to the taxing master that in his opinion the appointment of counsel had been justified by the scale or complexity of the dispute: and his omission of that phrase indicated the opposite opinion. In this way the taxing master was advised whether or not the costs of counsel should be included in his 'taxation'.

Under the 1996 Act the presumption of section 63 is that the arbitrator will exercise his power to 'determine by award the recoverable costs of the arbitration on such basis as [he] thinks fit'. Furthermore, section 65 empowers the arbitrator, unless otherwise agreed by the parties, to limit the recoverable costs of the arbitration to a specified amount. These provisions imply, in common with the Civil Procedure Rules 1998 (SI 1998 No 3132(L.17)), that the arbitrator will manage arbitration proceedings in accordance with those rules; in particular by ensuring that costs are not allowed to become disproportionate to the sum in dispute, and by himself determining the amount of costs to be paid by one party to the other.

The benefits resulting from the arbitrator determining the amount of recoverable costs are several. First, the arbitrator has a detailed knowledge of events during the interlocutory proceedings and at the hearing. He is therefore able to take account of the many considerations that may affect his determination, such as the need for legal representation and/or technical experts and whether it was reasonable to make such appointments. Second, he is already aware of events that may have led to costs being thrown away by delays or defaults on the part of either party. Third, he is already aware of any applications that may have been made by either party and of their outcome, for such matters as leave to amend statements of case, extensions of time for submission of such statements, and of any other matters which may have affected both the time taken for completion of the preliminaries and for the hearing, and the resulting costs.

The arbitrator's knowledge of factors affecting his determination of costs may be contrasted with the complete lack of knowledge of those considerations by a taxing master, who would have to obtain relevant information from the parties' advocates during what could become a long and costly 'taxation'.

### 8.4.11 Signature, seat and date

Subject to the right of the parties to agree otherwise, section 52(3) of the 1996 Act requires that 'The award shall be in writing signed by all the arbitrators...' Section 52(5) requires that 'The award shall state the seat of the arbitration and the date when the award is made.'

## 8.5 Notification of the award

Section 55 of the 1996 Act provides, subject to the right of the parties to agree otherwise, that the award is notified to the parties by service on them of copies of the award. However, provision is also made, under section 56(1), for the arbitrator to refuse to deliver his award to the parties except upon full payment of his fees and expenses.

In many cases the arbitrator may not have received full, or any, payment for his services by the time he makes his award. Generally the award provides that payment for it is the responsibility of the losing party, and having lost its case, that party is likely to be reluctant to pay for the award. For that reason an arbitrator who has not received full or any payment usually takes the precaution of notifying the parties of the availability of the award upon payment of his charges or any outstanding balance thereof by either party, and he provides in the award for reimbursement of those charges by the party responsible for paying them in the event that the other party pays them in taking up the award (see SD/19 in Appendix A).

Section 56(2) of the 1996 Act makes provision for a party faced with an arbitrator's refusal to deliver his award until he has been paid, to pay the sum demanded into court and to apply to the court for an order that the arbitrator deliver the award. The court may then determine the amount of such fees and expenses that are properly payable, pay that sum to the arbitrator, and repay to the applicant the balance, if any, of the sum paid into court.

## 8.6   *Correction of accidental errors*

Section 57(3) of the 1996 Act empowers the arbitrator, subject to agreement to the contrary by the parties, to correct an award

> '(a) ... so as to remove any clerical mistake or error arising from an accidental slip or omission or clarify or remove any ambiguity in the award.'

Applications for corrections pursuant to this section must be made within 28 days of the date of the award or such longer period as the parties may agree, and where the arbitrator invokes the provision on his own initiative, he must make the correction within 28 days of the date of the award, again with the proviso that the parties may agree to a longer period. The arbitrator is required, before exercising his power under the section, to afford the other parties a reasonable opportunity to make representations to him.

It is rarely necessary in practice to invoke this power, which corresponds with that available under the 'slip rule' in section 17 of the 1950 Act, because arbitrators should, and usually do, take great care to ensure that their awards do not contain mistakes or errors. Contention sometimes arises as to whether or not an error in the award falls within the definition 'any clerical mistake or error arising from an accidental slip or omission' (which wording is almost identical to that used in the 1950 Act). In *Mutual Shipping Corporation* v. *Bayshore Shipping Co* [1985] 1 WLR 625, it was held in the Court of Appeal that an arbitrator who mistakenly attributed evidence to the wrong party, causing him to make an award in favour of the wrong party, had power to correct that error under section 17 of the 1950 Act.

## 8.7 Additional awards

Section 57(3) of the 1996 Act also makes provision for the arbitrator, either on his own initiative or on the application of a party

'(b) ... to make an additional award in respect of any claim (including a claim for interest or costs) which was presented to the tribunal but was not dealt with in the award.'

A party's application for such an additional award must be made within 28 days of the date of the award: and the arbitrator's power to make such an additional award must not be exercised without affording the other parties (or impliedly all of the parties, where the arbitrator acts on his own initiative) to make representations to him. Any such additional award must be made within 56 days of the date of the award or such longer period as the parties may agree.

# CHAPTER NINE
# COSTS

## 9.1 Synopsis

The arbitrator has a discretionary power, under the 1996 Act and subject to the parties' right to agree otherwise, to award costs, to determine the amount of such costs, and to limit the amount of recoverable costs.

## 9.2 Statutory provisions

The 1996 Act includes the following provisions.

Under section 59, a definition of the term costs as comprising

(1) the arbitrator's fees and expenses
(2) the fees and expenses of any arbitral institution concerned and
(3) the legal or other costs of the parties.

Under section 60, an agreement as to payment by a party of all or part of the costs is valid only if entered into after the dispute has arisen.

Under section 61, power is vested in the arbitrator, subject to any agreement of the parties, to allocate the costs of the arbitration as between the parties. In doing so, again subject to the right of the parties to agree otherwise, the general principle to be followed is that 'costs follow the event' except where the arbitrator finds that rule to be inappropriate.

Under section 62, again subject to the right of the parties to agree otherwise, any agreement as to how costs are to be borne extends only to recoverable costs.

Section 63 empowers the parties to agree as to what costs are recoverable, and provides rules as to such costs to be applied in the absence of agreement. The fall-back provision, in the absence of agreement, empowers the arbitrator to determine the recoverable costs. Failing such determination, any party may apply to the court for determination.

111

Section 64 limits the recoverable costs of the arbitrator, subject to the parties' right to agree otherwise, to 'such reasonable fees and expenses as are appropriate in the circumstances'.

Finally, section 65 introduces a provision, subject to the right of the parties to agree otherwise, empowering the arbitrator to limit the amount of recoverable costs to a specified amount. Any such limit, or variation of such a limit, may be made at any stage, but it must be sufficiently in advance of incurring the costs to which it relates to enable the parties to take it into account.

## 9.3 Basis of award of costs

The basic principle under which responsibility for costs is to be determined is contained in section 61 of the 1996 Act which, somewhat surprisingly in an Act which generally avoids the use of legal jargon, quotes the rule that 'costs follow the event', meaning that the successful party is entitled to an award of costs. No difficulty arises in applying this rule to a simple case, where a claim either succeeds or fails in its entirety, and where there is no default on the part of the successful party. In practice however, complications usually arise: the claim may succeed in part, there may be a counterclaim which succeeds wholly or in part, the respondent may have made an offer which the claimant ought to have accepted, or costs may have been incurred unnecessarily by the successful party. Any of these events may warrant an adjustment to the basic rule.

Where a party succeeds in a substantial part of its claim, the fact that it does not wholly succeed is no reason for failing to award costs to that party. Arbitrators are often criticised for their tendency to apportion costs in such circumstances, for example to award one half of the claimant's costs where one half of its claims succeed. The correct principle is to recognise that the claimant had to invoke arbitration in order to obtain an award of money to which it was entitled; and that it was open to the respondent to make an offer in settlement equal to or greater than the sum awarded (see paragraph 9.4 below) in order to protect it against an award of costs.

In *Channel Island Ferries Ltd* v. *Cenargo Navigation Ltd (The* Rozel) (1994) 2 Lloyds Rep 161, Phillips J dismissed an appeal from an arbitrator's award to the claimant of the entirety of his costs where the arbitrator had awarded significantly less than the sum claimed. The learned judge commented that:

'...counsel for the owners had suggested that where a claimant had recovered significantly less than claimed, the practice of not

112

awarding him the entirety of his costs, on the ground that he had been only partially successful, had become prevalent among arbitrators. If there was such a practice, it was desirable that it should stop. Gross exaggeration of a claim, or the expenditure of substantial time on discrete issues of fact on which a claimant was unsuccessful could justify an award of less than all of a successful claimant's costs, especially if a claimant's conduct had, in the arbitrator's opinion, been unreasonable. In the instant case, the arbitrator had had in mind a large number of discrete items of claim some of which had succeeded and some of which had failed. His lordship would dismiss the owner's appeal.'

Where there is a claim and a counterclaim, both of which succeed in whole or in part, the general rule is that the nett winner should be awarded its costs; for the reasons referred to above. In *Tramountana Armadora* v. *Atlantic Shipping Co* [1978] 1 Lloyds Rep 391, Donaldson J (as he then was) set down the principles to be applied generally in the award of costs. In dealing with a claim/counterclaim situation, he stated

'...On some occasions it is then appropriate to consider each separately and, for example, to give the claimants the costs of the claim and the respondents the costs of the counterclaim. This leaves it to the parties to agree, or to the taxing authority to determine, what proportion of the costs of each party is attributable to the claim and what to the counterclaim. On other occasions it may be clear to the judge or arbitrator that the claim and counterclaim have no separate existence ... In such a case it is usually inappropriate to make cross orders for costs.'

The former situation is unusual in construction contract disputes, in which a counterclaim usually arises as a defence to a claim, and is based upon the same facts. The judgment in the *Tramountana* case also emphasises the principle that the discretion as to the award of costs lies with the arbitrator and not with the court. While the court would give directions as to the principles to be adopted by the arbitrator in the exercise of his discretion it would not and could not exercise that discretion on the arbitrator's behalf.

Where one party makes an offer to settle the claims in a sum sufficient to cover the amount ultimately awarded by the arbitrator, plus costs incurred by the offeree up to the date of the offer, and that offer is refused, then responsibility for the continuation of the dispute rests upon the offeree, who should be ordered to bear costs incurred after the date of the offer. In assessing the adequacy of any such offer the arbitrator should take into account interest to which the offeree would have been entitled as at the date of the offer, and he should of course also ensure that in comparing the offer with his own assessment of the claims he compares like with like. Hence, for example, no allowance need be made for costs incurred up to the date of the offer where the offer is expressed as being '£x plus costs'.

Where the successful party is responsible for having incurred costs unnecessarily – for example by failing to appear at a meeting which then has to be adjourned, or by applying for leave to amend its statement of case at a late stage, making it necessary to vacate dates reserved for the hearing – that party should be required to bear all costs occasioned by its default. Such a requirement is usually confirmed at the time of the default, by an order that the defaulting party bear such costs 'in any event', that is, irrespective of the outcome of the arbitration and the arbitrator may direct that such default costs be paid by the defaulting party immediately. Where an arbitrator makes an award of costs that does not comply with the basic rule that costs follow the event, he should explain within his award his reasons for doing so.

## 9.4   Offers to settle

It will be seen from the above that the making of an offer of sufficient magnitude to warrant its acceptance protects the offeror against an award of costs incurred after the date of the offer. The format of such an offer, and the machinery by which it is made to the opposing party and, at the appropriate time, brought to the notice of the arbitrator, needs to be considered.

### 9.4.1   Open offers

Either party may make an offer to settle, and may notify that offer to its opponent in an open letter, that is, a letter not marked 'without prejudice' or 'without prejudice save as to costs'. In the absence of such marking the letter may, if the arbitration proceeds, be adduced as evidence of admission of liability, thereby detracting from the credibility of the offeror's submissions denying liability. It is for that reason that an offer to settle is more usually made in the form of a Part 36 offer.

### 9.4.2 Part 36 offers

Formerly known as a *Calderbank* letter, an offer to settle the whole or part of a claim is, under the Civil Procedure Rules 1998 (CPR 98) termed a Part 36 offer. It is subject to the provisions of those rules, which require that the offer

- must be in writing;
- may relate to the whole claim or part of it or any issue and must define its ambit;
- must state whether it takes account of any counterclaim;
- must state that it is inclusive of interest, or state whether interest is offered and if so the amount or rates and period for which it is offered;
- must state that it is open for acceptance within 21 days, after which the offeree may accept only if the parties agree liability or the court gives permission.

In general the CPR 98 relate mainly to litigation, and the rules require that where it is possible for a defendant to make a Part 36 payment (which corresponds to a payment into court in earlier terminology) a Part 36 offer is not effective.

Because in arbitration there is no court into which a payment could be made, Part 36 offers are in principle valid. It is however necessary to consider the machinery by which the arbitrator becomes aware, after he has made his substantive award, of the existence and amount of such an offer, in order that he may take it into consideration in his award of costs. There are two methods in general use.

Under the 'sealed offer' method, a copy of the Part 36 offer, which is made in a letter marked 'without prejudice save as to costs' is put into an envelope, sealed, and handed to the arbitrator at the conclusion of the hearing, with a request that it is not opened until after the arbitrator has made his substantive award. At that stage, before dealing with costs, the arbitrator opens the envelope and ascertains whether or not the offer contained in the letter is such as to have warranted its acceptance at the date when it was made. If the answer to that question is in the affirmative, then the arbitrator awards all costs incurred subsequently to the date of the offer (or a week or two later in order to allow reasonable time for consideration of the offer) to the offeror, on the basis that the offeree was responsible for continuation of the dispute after having received an acceptable offer.

One of the defects of the sealed offer procedure is that the arbitrator does not need to be clairvoyant in order to know that the envelope handed to him contains a copy of an offer. That defect may easily be avoided if the arbitrator requires the respondent (or, where there is a counterclaim, each party) to hand to him a sealed envelope which contains either a copy of any Part 36 offer that may have been issued or a statement that no such letter has been issued. The second defect is not so easily resolved. Unlike a payment into court, the sealed offer is not backed up with money: there is no guarantee that the offered payment would have been made had the offer been accepted. It would of course be open to the offeror to open a dedicated account at a bank, and to obtain the bank's confirmation that the amount of the offer is available for payment when authority to make such payment is received. Such a procedure does not however appear to be in general use.

An alternative to the sealed offer procedure is for the arbitrator to indicate, usually at the preliminary meeting, his intention to make and publish his first award covering matters other than costs, and thereafter to invite the parties to address him, either orally or in writing as may be agreed, before he makes his award of costs. At that stage the existence of any offers made may be brought to light by the parties.

Where the arbitrator does not himself suggest such a procedure, either party may ask that he provides an opportunity for them to address him on costs before he makes his award on that matter.

## 9.5  Failure to award costs

Subject to the right of the parties to agree on the powers of the arbitrator to correct an award or to make an additional award, section 57(3)(b) of the 1996 Act provides that, where the arbitrator fails to deal in his award with costs the arbitrator may, either on his own initiative or on the application of a party

'make an additional award in respect of any claim (including a claim for interest or costs) which was presented to the arbitrator but was not dealt with in the award.'

Any such application must be made within 28 days of the date of the award or such longer period as the parties may agree.

116

## 9.6 Determination of recoverable costs

Section 63(1) of the 1996 Act empowers the parties to agree what costs of the arbitration are recoverable, while section 63(3) empowers the arbitrator, in the absence of such agreement, to determine the recoverable costs on such basis as he thinks fit. He must however specify in his award the basis on which he has acted, the items of recoverable costs, and the amounts referable to each. Where the arbitrator does not determine the recoverable costs any party to the proceedings may apply to the court for determination of recoverable costs *on such basis as it thinks fit*. The basis of determination, unless otherwise directed by the arbitrator or by the court, is, under section 63(5) of the 1996 Act

'(a)    ... there shall be allowed a reasonable amount in respect of all costs reasonably incurred, and

(b)    any doubt as to whether costs were reasonably incurred or were reasonable in amount shall be resolved in favour of the paying party.'

The basis of determination of costs under section 63(5) of the 1996 Act is substantially the same as that in rule 44.4(2) of the Civil Procedure Rules 1998 where costs are awarded on the *standard* basis. Section 63(3) of the Act however implies that the arbitrator may determine costs on another basis, and rule 44.4(1) of the CPR provides for assessment either on the standard basis or on the *indemnity* basis.

It is unusual for an arbitrator to determine costs on any basis other than the standard basis. Where for some special reason the arbitrator deems it appropriate to do so, that reason should be stated, and the arbitrator should observe the definition of the indemnity basis given in the CPR: '...the court will resolve any doubt which it may have as to whether costs were reasonably incurred or were reasonable in amount in favour of the receiving party'.

It is suggested that the arbitrator should, at an early stage in the proceedings, indicate his intention to determine the recoverable costs of the arbitration, in exercise of his powers under section 63 of the 1996 Act. Most arbitrators find no difficulty in determining the amount of their own costs, having, where possible, agreed with the parties as to the hourly and/or daily rates to be charged, and they know how much time they have devoted to the proceedings and the amount of expenses they have incurred. Hence there is no logical reason why arbitrators should seek outside assistance in determining the amount of their own charges.

Similar considerations apply to the determination of the parties' costs; but arbitrators are often reluctant to do so. That reluctance is, it is submitted, illogical. As compared with a taxing master in the court, the arbitrator has many advantages when determining the parties' costs. He is already familiar with details of the contract, of the disputed matters that arose from it, and with the manner in which those matters were dealt with in the arbitration. He knows whether or not the parties chose appropriate advocates to present their cases: whether any appointment of experts was necessary: whether either party was obstructive or dilatory in the course of the proceedings or wasted time in unnecessary challenges to the arbitrator's jurisdiction. He is also likely to be able to carry out his determination of the costs far more rapidly than a taxing master in the court. It is therefore suggested that, except where the parties agree to the contrary, the arbitrator should exercise his power under section 63(3) to determine the parties' costs; and should state his intention to do so in his award of costs, using the phrase: 'such costs, if not agreed, to be determined by me'.

Having so directed the arbitrator should, if and when the parties find they are unable to agree, invite the party to whom he has awarded costs to submit its bill of costs; he should invite the paying party to submit its objections to that bill and he should invite the receiving party to reply to those objections. Thereafter if the parties so agree it may be possible for the arbitrator to make his determination solely upon the written submissions, although if either party so requires he should provide an opportunity for oral submissions. It is also suggested that a simple statement of the time spent by each of the receiving party's advocates, staff, experts and counsel, and of their hourly rates, is sufficient to enable the arbitrator to make a reasonable determination in accordance with the basis quoted above.

## 9.7 *Arbitrator's fees and expenses*

The terms 'costs of the reference' and 'costs of the award' as used in the 1950 Act have been superseded in section 59(1) of the 1996 Act, no doubt in line with the policy of avoiding legal jargon. The latter term is now simply 'the arbitrators' fees and expenses', and the former has become 'the legal or other costs of the parties', both of which are self-explanatory. A third category of costs has been introduced; namely 'the fees and expenses of any arbitral institution concerned', the need for which may not arise, because such costs will generally have been paid by one or other of the parties, and could therefore be included in that party's costs. The principles referred to above by which costs are awarded and are determined apply to each of the three types of costs.

## 9.8    *Interest on costs*

The arbitrator's discretionary power under section 49 of the 1996 Act to award 'simple or compound interest from such dates, at such rates and with such rests as it considers meets the justice of the case ... and from the date of the award on the outstanding amount of any award (including ... any award as to costs)' should be exercised judicially; that is, in accordance with established principles of law. In *Hunt* v. *R.M. Douglas (Roofing) Ltd* [1988] 3 WLR 975, the House of Lords overruled an earlier decision that interest was payable from the date of the taxing master's certificate, holding that interest is payable from the date when judgment is pronounced.

That decision, it is submitted, is entirely consistent with the arbitrator's discretionary power under section 49(4), and it follows that, unless there is some valid reason to the contrary, the arbitrator should, where he makes an award of costs, also award interest on such costs as from the date on which they were incurred.

# CHAPTER TEN
# POWERS OF THE COURT IN RELATION TO THE AWARD

## 10.1 Synopsis

The powers of the court in relation to the award include enforcement, and determination of challenges to the award on any of the several bases that may be available to a dissatisfied party.

## 10.2 Enforcement

Where a losing party fails to honour its obligations under an award the other party may apply to the court under section 66 of the 1996 Act for leave (now referred to under CPR 98 as 'permission') to enforce the award in the same manner as a judgment or order of the court to the same effect. Where such leave is given, judgment may be entered in the terms of the award.

The applicant need not however, having obtained the court's permission, actually enter the judgment, but may execute the award as if it were a judgment. Enforcement measures include, where the award is for payment of a sum of money, the seizure of the respondent's goods, or in a case where a court order of payment is not complied with, committal of the respondent to prison for contempt.

A possible ground on which an objection may be raised to such an application is that the arbitrator lacked substantive jurisdiction to make the award. An objection on that basis must however, under section 73 of the 1996 Act, be made forthwith or within such time as is allowed by the arbitration agreement or by the arbitrator. A party who takes part, or continues to take part, in the proceedings and fails to object may not raise that objection later, unless he shows that when he did so he did not know and could not reasonably have discovered the grounds for the objection.

The power of enforcement of awards made under other enactments or rules of law is unaffected by this provision of the 1996 Act.

Section 99 of the 1996 Act provides that Part II of the 1950 Act (see Appendix C), which deals with the enforcement of foreign awards under the Geneva Convention which are not also New York Convention awards, remains in force. Section 100 of the 1996 Act similarly provides for the recognition and enforcement of New York Convention awards.

## 10.3 Challenges to the award

There are three bases upon which arbitrators' awards may be challenged: first that the arbitrator lacked substantive jurisdiction, second that there was a serious irregularity affecting the arbitrator, the proceedings, or the award; and third on a question of law. Applications or appeals to the court may not, under section 70 of the 1996 Act, be brought unless the applicant or appellant has first exhausted any available process of appeal or review, and any available recourse under section 57 of the Act, which empowers the arbitrator to correct an award or to make an additional award in respect of a claim which was presented to the arbitrator but not dealt with in the award.

### 10.3.1 Right to object

Most challenges to the arbitrator's jurisdiction arise at an early stage in the proceedings, see Chapter 4. Similarly, allegations of improper conduct of the proceedings, failure to comply with the arbitration agreement or with the 1996 Act, or other irregularity, usually relate to the proceedings rather than to the award. Section 73 of the 1996 Act provides that a party who takes part in the proceedings without making an objection on any of those grounds forfeits his right to do so unless he can show that at the time he took part or continued to take part in the proceedings, he did not know and could not have known the grounds for his objection. That section should be effective in debarring disgruntled losers from challenging procedures when they become aware that the award is not in their favour.

### 10.3.2 Serious irregularity

Section 68 of the 1996 Act provides a right of a party to challenge an award on the grounds of 'serious irregularity' affecting the

121

arbitrator, the proceedings or the award. Serious irregularity is defined in section 68(2) as covering nine different types of what in the 1950 Act was termed 'misconduct of the arbitrator or of the proceedings'. The main element of any such irregularity is failure to act fairly, but there are also several other faults falling within the definition, such as the arbitrator's failure to comply with the agreed procedure, exceeding his powers, failure to deal with all of the issues put to him, uncertainty or ambiguity as to the effect of his award, obtaining the award by fraud or in a way contrary to public policy, and failure to comply with requirements as to the form of the award. Finally there is a blanket type of irregularity defined as being

> 'any irregularity in the conduct of the proceedings or in the award which is admitted by the arbitrator or by any arbitral or other institution or person vested by the parties with powers in relation to the proceedings or the award.'

The powers available to the court under section 68(3) of the 1996 Act, in dealing with serious irregularity affecting the arbitrator, the proceedings or the award are

> '(a)  to remit the award to the arbitrator, in whole or in part, for reconsideration;
> (b)  to set the award aside in whole or in part; or
> (c)  to declare the award to be of no effect, in whole or in part.'

There is a proviso however, requiring that the power in either (b) or (c) above shall not be exercised unless the court is satisfied that it would be inappropriate to remit the award.

Under section 23(1) of the Arbitration Act 1950, similar provision was made for removal by the High Court of an arbitrator who 'has misconducted himself or the proceedings', while under section 24(1) of that Act provision was made for the High Court to revoke the authority of an arbitrator named in an agreement on the ground that the arbitrator so named 'is not or may not be impartial'. In *R* v. *Gough* [1993] AC 646, it was held by Lord Goff in the House of Lords that 'the test should be the same in all cases of apparent bias, whether concerned with justices or members of other inferior tribunals, or with jurors, or with arbitrators', and that test was:

> '... whether ... there was a real danger of bias on the part of the relevant member of the tribunal, in the sense that he might

unfairly regard (or have unfairly regarded) with favour or disfavour, the case of a party to the issue under consideration by him.'

In *Brian Andrews* v. *John H. Bradshaw and H. Randell & Son Limited* [2000] BLR 6 the Court of Appeal allowed an appeal by the arbitrator from the judgment of Judge Brian Knight QC, sitting on business and commercial lists of the Central London County Court, removing the arbitrator pursuant to an application under section 24(1) of the 1996 Act. The Court of Appeal held that the test laid down by Lord Goff in *R* v. *Gough* was applicable to allegations of bias under section 24(1) of the 1996 Act. Lord Justice Mance added that 'When speaking of real danger, the court is thinking in terms of possibility rather than probability of bias'.

### 10.3.3   Errors of law

Section 69 of the 1996 Act provides a limited right of appeal to the court on a question of law arising out of an award. The procedure does not differ substantially from that incorporated in the 1979 Act except in respect of exclusion agreements (see below); but it is expressed in terms that are more easily understood, especially by those who are unfamiliar with legal terminology.

It is open to the parties to agree, either before or after a dispute has arisen, to exclude the provision for appeals on questions of law. Such an agreement is termed an 'exclusion agreement' and an agreement to dispense with reasons in the award is considered under the 1996 Act to constitute such an agreement. Provision is made, in section 87 of the 1996 Act, for exclusion agreements to be ineffective, in the case of domestic arbitrations, unless entered into after the dispute has arisen, which provision would bring the 1996 Act into line with that under section 4 of the 1979 Act. Section 87 of the 1996 Act has not however been brought into operation, and is likely to be repealed. The parties are therefore free to enter into a binding exclusion agreement either before or after their dispute arises. There are however a number of other conditions that have to be satisfied before leave to appeal is given.

First, an appeal shall not be brought except either with the agreement of all of the parties to the proceedings, or with the leave of the court. In general it is unlikely that a successful party would be willing to agree to an appeal against an award that is in his favour; and therefore the usual need is to obtain the leave of the court.

Before giving such leave the court requires to be satisfied under section 69(3)

(a)  that determination of the question will substantially affect the rights of one or more of the parties;

(b)  that the question is one which the arbitrator was asked to determine;

(c)  that, on the basis of the findings of fact in the award, either the decision of the arbitrator is obviously wrong, or the question is one of general public importance and the decision of the arbitrator is at least open to serious doubt; and

(d)  that, despite the agreement of the parties to resolve the matter by arbitration, it is just and proper in all the circumstances for the court to determine the question.

Where an application is made for leave to appeal under section 69 of the 1996 Act the applicant must identify the question of law to be determined and state the grounds on which leave should be granted. The court may at its discretion determine the application for leave to appeal with or without a hearing; and leave of the court is also required for an appeal against the court's decision to grant or to refuse leave to appeal.

Where leave to appeal is given, the court is empowered to confirm, vary, remit to the arbitrator in whole or in part, or to set aside the award in whole or in part. The power to set aside must not however be exercised unless the court is satisfied that it would be inappropriate to remit the matters in question to the arbitrator for reconsideration.

Section 69(6) of the 1996 Act provides that

'The leave of the court is required for any appeal from a decision of the court under this section to grant or refuse leave to appeal.'

In *Henry Boot Construction (UK) Ltd* v. *Malmaison Hotel (Manchester) Ltd* (*The Times*, 31 August 2000) it was held by the Court of Appeal that it did not have jurisdiction to grant permission itself, nor to review the refusal of the High Court, or county court, to grant permission.

# CHAPTER ELEVEN
# DISPUTE AVOIDANCE AND MANAGEMENT

## 11.1  Synopsis

Probably the most important aspect of dispute management is the avoidance of disputes; which however resolved usually involve the payment of substantial costs by one party or the other. This chapter covers, chronologically, management procedures designed to minimise the risk that disputes may arise, and where they do arise, the losses resulting from them, which procedures commence when a construction project is conceived and end when the final payment is made.

## 11.2  Contract documents

### 11.2.1  Adequacy of site investigation

Disputes frequently arise from clause 12 of the ICE Conditions of Contract, under which the contractor is entitled to reimbursement of additional costs reasonably incurred, together with a reasonable percentage addition thereto in respect of profit, in dealing with physical conditions or artificial obstructions which could not reasonably have been foreseen by an experienced contractor. Such costs may cover both the costs of additional work made necessary by the conditions or obstructions, and the costs of delay and disruption to the contractor's programme of work. From the employer's viewpoint the consequences of a successful claim under clause 12 may be especially damaging, for the following reasons.

- The cost of the additional work is evaluated on a *cost plus profit* basis and is therefore likely to be much greater than would have been incurred had the work been included in a competitive tender.

- The additional costs of delay and disruption to the work may be substantial.
- Being unforeseeable, the costs could not have been allowed for in the employer's budgetary provisions.

Where a thorough site investigation is made the likelihood that unforeseen conditions or obstructions will be encountered is eliminated or at least reduced. The engineer for the project is able to base his design upon known ground conditions, to ensure that necessary work is included in the contract, and hence to obtain competitive prices for that work. For example, the need to base a structure on piled foundations, foreseen and allowed for in the contract, enables competitive prices to be obtained for that work, and enables the contractor to plan his work incorporating the piling, often by a subcontractor. Where that need is not foreseen delay is often caused by the abandonment of the original design, the need for additional boreholes, the need to redesign the foundations, the need to negotiate a subcontract for the varied work, and by the execution of the work. Besides delaying the project these variations result in additional cost, not only of the piling itself, but also of the associated works and the delay and disruption to the contractor's programme.

## 11.2.2 Accuracy of quantities

Bills of quantities incorporated in the contract should include accurate quantities of the measured work expected to be required, any adjustments being in the form of rounding off figures on a give-and-take basis. That objective is not always achieved, sometimes because arithmetical errors are made, and sometimes because of deliberate distortion of the true quantities.

A common ploy of some engineers is to include an excessive quantity in an item in which the quantity of work is not easily measured at tender stage, such as an 'extra over' item for rock excavation, with the objective of providing a reserve of funds to offset additional costs that may arise elsewhere. Again, where the engineer is anxious to ensure that tenders submitted are within his budgetary prediction, he may be tempted to enter an artificially low quantity against some items, especially where the cost of the work is high and the final quantities are difficult to predict, such as pre-blasting in preparation for dredging. Any such distortion of quantities may have adverse consequences for the employer.

A tenderer who suspects that the quantity entered against an item in the bill of quantities is incorrect may, and sometimes does, seek to take advantage of that situation, by inflating his rates for items in which he expects that the billed quantity will be exceeded, and reducing his rates where he expects a reduction in the billed quantity. Adjustments may need to be made elsewhere in the tender to compensate for the distorted rates; but this is not usually a difficulty. The nett result, if the contractor's predictions are correct, may be a substantial increase in the cost of the project. The engineer should safeguard his client against such tactics by ensuring that the quantities included in the bill are accurate. It is to be expected that at tender stage the engineer should have a better knowledge than tenderers of the site conditions and hence of the quantities of work that are likely to be required.

In *Henry Boot Construction Ltd* v. *Alston Combined Cycles Ltd* [2000] BLR 247, a question arose as to whether or not an erroneous rate in a contract under the ICE conditions, sixth edition, ought to be applied to a variation involving similar work to that in the contract. Clause 52(1) of the conditions provides three rules for the evaluation of varied work.

- 'Where work is of similar character and executed under similar conditions to work priced in the Bill of Quantities it shall be valued at such rates and prices contained therein as may be applicable.
- Where work is not of a similar character or is not executed under similar conditions or is ordered during the Defects Correction Period the rates and prices in the Bill of Quantities shall be used as the basis for valuation so far as may be reasonable
- failing which a fair valuation shall be made.'

The arbitrator, John Tackaberry QC, had held that it would not be 'reasonable' to apply a rate in the contract known to contain at least one mistake, under the second rule in clause 52(1). Judge Humphrey LLoyd QC allowed an appeal against the award, and remitted it to the arbitrator with a direction that the contract rate should be applied. On appeal from that judgment the Court of Appeal upheld, by a majority, the learned judge's decision, remitting the award to the arbitrator.

### 11.2.3 Clarity of contract documents

Many construction contract disputes originate from variations to standard forms of contract documents, which may be incorporated deliberately or inadvertently. It is not unknown for an employer or an engineer to introduce varied or additional clauses into a standard form of contract in order to avoid some potential liability, such as interest on overdue payments. Where such variations or additions are in conflict with the standard form it may be arguable that the variation should take precedence over the unamended form. *Keating on Building Contracts* sixth edition at page 45 quotes the rule:

> '...the written words are entitled to have a greater effect attributed to them than the printed words, inasmuch as the written words were the immediate language and terms selected by the parties themselves for the expression of their meaning.'

The principle is of course extended to cover not only handwritten words, it also covers typed words, which have greater effect than the standard printed words. It is submitted that the employer, and his engineer, should avoid the introduction of additional or varied provisions in a contract except where it is essential to do so, and that any such provisions should, in the interests of both parties, be clear and unambiguous, and should deal expressly with any conflict between them and standard clauses.

# 11.3   Basic contract law

## 11.3.1   The tender

A contract is created by a valid acceptance of a valid offer. Certain other elements must also be present: there must be consideration, an intention to create a legally binding relationship, and certainty; but these are usually adequately covered in the offer, which in construction parlance is the tender.

Whether or not the tender documents issued by the employer so provide, the tenderer may limit the validity of his tender to a defined period of time. If by so doing he contravenes a requirement of the employer then the employer may waive that requirement, negotiate with the tenderer for its removal, or ignore the tender. Additionally, and notwithstanding any undertaking he may have given, the tenderer may withdraw the tender at any time *prior to its acceptance*. If the tenderer neither limits the period of validity of his tender nor withdraws it, then it remains open for acceptance within a reasonable period of time: the definition of 'reasonable' being a matter for determination by the court.

Alternatively it is arguable that the correct forum for the determination of a dispute as to the reasonableness of the period of time that elapsed between submission of the tender and its purported acceptance is arbitration, notwithstanding the fact that the existence of the contract containing the arbitration agreement is the subject of the dispute. This is because section 7 of the 1996 Act provides that an arbitration agreement that was intended to form part of another agreement is separable from that other agreement and remains effective even if the other agreement did not come into existence.

Where the tenderer receives a reminder, in the form of an acceptance long after it was submitted, that his tender remains open, he may, if he no longer wishes to perform the work, challenge the validity of the acceptance on the ground that it has not been issued within a reasonable period of time. It is too late, at that stage, to demand an increase in the amount of the tender to take account of inflation or any other factor, although such an increase could be agreed by negotiation where the employer concedes that the acceptance may have been unduly late. Again, the tenderer cannot, if he proceeds with the work, expect to substantiate a claim in respect of the lateness of the acceptance.

In preparing the tender the contractor should remember that should a dispute arise as to the value of varied work, the basis upon which the contract work was priced is likely to be relevant. Pricing notes may have to be disclosed; hence such notes should be clear, logical and legible. This applies especially where a lump sum cost, such as that of shuttering to be used many times, is spread over the number of uses in a rate for measured work, and therefore the rate needs to be adjusted if the number of uses is varied. Similarly an adjustment may be needed in the rate for a commodity, such as filling, where a limited supply is available. An increase in the quantity may incur a need to seek elsewhere for the additional material, and to bear additional haulage costs, for which an upward rate adjustment may be claimable.

The contractor should also ensure that his methods of construction, plant and programmes of work are clearly defined, not only for submission to the engineer for approval in accordance with the terms of the contract, but additionally to substantiate claims for delay or disruption. Where necessary the contractor may need to submit his work programme with his tender, and make it clear that the tender is conditional upon the acceptance of the programme. In *Glenlion Construction* v. *The Guinness Trust* (1987) 39 BLR 89 it was held in the Official Referee's court that 'there was no implied term of the contract ... that if and in so far as the programme showed a completion date before the [contractual] date for completion the employer ... his servants or agents should so perform the said agreement as to enable the contractor to carry out the works in accordance with the programme and to complete the works on the said completion date.'

It follows from that decision that a contractor cannot rely upon the engineer's or the employer's cooperation in achieving a completion date earlier than that prescribed by the contract, hence that the contractual date for completion must be altered if the contractor seeks to reduce his site oncosts by completing in a shorter period than is provided for in the contract. A consequence of such an alteration is that the contractor is then bound by the earlier date, and may be liable for liquidated damages should he fail to complete by that date.

## 11.3.2 The acceptance

The second element needed to create a contract is a valid acceptance: that is, an acceptance during the period of validity of the offer, and in terms compatible with it. A simple statement 'I hereby accept your offer' is sufficient, provided of course that sufficient information is given to identify the offer that is being accepted. If the acceptance is not in terms compatible with the offer, or is qualified in any way, then it is not an acceptance but a *counter-offer*, which itself requires acceptance before a contract is created. For example, a statement that a qualified tender is accepted subject to the removal of the qualification is a counter-offer, which the tenderer may accept or may possibly accept subject to an increase in the tender price, thereby making a further counter-offer. It is only when an offer or counter-offer is accepted without qualification that a contract is created.

Both parties should however be aware that an acceptance may be deemed to have been given by a party's action, for example in starting work upon receipt of a qualified acceptance. Such action will generally constitute acceptance of the qualification, and thereby create a contract. Similarly the employer's instruction to commence work will generally constitute an acceptance of the tender, with whatever qualifications may be contained in it. The not uncommon practice of giving such an instruction before contractual details are finally settled may result in the employer being deemed to have accepted the last offer made.

The acceptance – and indeed the tender – need not be in writing in order to be valid. An oral contract is just as valid and enforceable as a written contract, but it does have disadvantages in that it is more difficult to prove both its existence and its terms.

### 11.3.3  Letters of intent

The usual purpose of a letter of intent is to enable preliminary work to proceed while formalities such as the obtaining of financial or planning approvals are completed. Where preliminary work (such as the design of temporary works, the ordering of materials, or the erection of site offices) is carried out on the authority of such a letter, no difficulty arises if the approvals are obtained and the contract comes into being. Problems may however arise where for some reason the contract does not materialise: a situation that may not have been contemplated by the parties.

The nature of the problem depends upon the wording of the letter of intent. It is to be expected that the intention of the parties is that the letter does not constitute an acceptance of the tender, but does give limited authority to the tenderer to proceed with certain defined preliminary works in advance of that acceptance, on the basis that should the contract fail to come into existence, the tenderer will be recompensed for any of the authorised work carried out. Where this situation arises, materials obtained and paid for under such an arrangement will of course become the property of the employer: and the tenderer will be responsible for the safe custody of such materials until the date of handing over. Where the contract comes into existence, payments made against such costs are credited to the employer's account, and the materials are in due course built into the works in accordance with the contract.

Unsatisfactory wording of a letter of intent may result in disputes of various kinds. First, it may be possible for the tenderer to argue that the letter of intent, and possibly other actions of the employer, constitute an acceptance of his tender, hence that the employer is bound by the terms of the contract. More usually the tenderer may find that he is not reimbursed for expenditure on preliminary work which he understood to have been authorised, and is unable to substantiate his right to payment. An example of a letter of intent drafted with the objective of avoiding possible misunderstandings is provided at SD/21 in Appendix A.

### 11.3.4  The formal agreement

The contract having been created by a valid acceptance of a valid offer, no further documentation is necessary to ensure that it is enforceable by either party. The ICE Conditions, for example, include in the Form of Tender the statement 'Unless and until a formal Agreement is prepared and executed this Tender together with your written acceptance thereof, shall constitute a binding Contract between us', and that is of course a simple statement of the legal position. The ICE Conditions, and other standard forms of contract, do however provide in addition for the signing – and sometimes sealing – of a formal agreement. Such a document is of little significance except in that, where executed under seal, it extends the period under the Limitation Acts during which an action may be brought, from six to twelve years. It may however more generally be useful where negotiations have preceded the making of the contract, in that it may be used to incorporate a summary of the outcome of the negotiations. Care is needed to ensure that any such summary accurately reflects the true outcome of the negotiations, because the formal agreement is deemed to supersede the negotiations that led to its preparation. An alternative form sometimes adopted is to incorporate in the formal agreement the correspondence between the parties during the negotiations leading to the agreement; which procedure has the merit of providing a clear definition of the documents included in the contract.

## 11.4  *The construction period: records*

Probably the most important factor enabling a contractor to substantiate claims arising under a construction contract or from breaches of it, and enabling the employer and his engineer to defend such claims, is the maintenance of adequate records during the construction period. Few civil engineering or building contracts are completed exactly as envisaged at the time of making the contract. In many cases there are variations; delays resulting from a variety of causes including variations and unforeseen ground conditions; and alterations in the employer's requirements. Variations may be needed in order to accommodate or take advantage of developments in technology, or to correct design errors; and such variations frequently cause delays or disruption of the contractor's programme of work. It is only by maintaining full and accurate records of such events as they occur that a contractor may be able to obtain adequate compensation for the costs and delays he incurs, and the employer may be able to ensure that he pays no more than the sum to which the contractor is properly entitled. In most civil engineering contracts the contractor's records should include the following.

- Correspondence with the engineer, the resident engineer, the employer, subcontractors, suppliers, and any other persons or organisations involved in the works. In many cases separate files may be needed for correspondence with each of the external organisations concerned, and it may be necessary to maintain separate files for variation orders, monthly valuations, engineers' certificates, notices of claims, minutes of progress meetings, and so on.
- Quotations and invoices received from subcontractors, suppliers, plant hirers and any other organisations entitled to payment.
- Site diaries recording the weather, the state of progress of each part of the works, and the fact and causes of delay to any part of the works.
- Progress charts, which in the case of underground works may include records of subsoil strata, ground water levels, etc.
- Copies of all programmes of work, including a dated copy of each revision of the original ICE clause 14 programme, or corresponding contractual programme.
- Details of all plant and labour employed on the site and also where and cost information relating thereto, allocated to the various parts of the project.
- Details of site oncosts, including supervisory and non-productive labour, offices, workshops, stores, mess-rooms, sanitation, water supply, site transport, small tools, insurances.
- Progress photographs, preferably taken from a series of fixed locations, looking in a fixed direction, in order that the month's work may be clearly illustrated. Each photograph should be marked with its location, direction of view and date.

The contractor should also ensure that records are kept of the date of issue of all revised drawings, and that such drawings issued during the progress of the works may be clearly distinguished from those issued for tendering purposes. Many of the above records should be reflected in similar records in the engineer's and the resident engineer's offices.

## 11.5 Claims

Where events occur that appear to warrant the submission of a claim the contractor should ensure that the following actions are taken.

- Record all relevant facts relating to the circumstances of the claim, including its cause, the instructions or lack of instructions from which it originates, whether written or oral, by whom given, dates and times of events.
- Where necessary confirm in writing any oral instruction from which the claim arises, pursuant to clause 2(6) of the ICE conditions, and ensure that the instruction given falls within the authority of its author.
- Identify the clause(s) of the contract under which the claim arises.
- Where the claim arises from a breach of contract, identify the clause(s) breached.
- Give notices pursuant to clause 53 of the ICE conditions and pursuant to the clause under which the claim arises, or corresponding provisions of other forms of contract. Where necessary give such information as is available at the time of notification of the claim, and undertake to provide further information as and when it becomes available.
- Where the claim arises from a breach of contract, give notice of the likely consequences of the breach, in terms of delay and extra cost.
- Ensure that all necessary evidence is available to substantiate the claim should it be disputed. Photographs of work, samples of materials encountered during excavations, or expert reports upon geological conditions may sometimes be needed, especially where work is to be covered up.

The contractor should at the earliest opportunity evaluate the claim and apply for payment as part of the next monthly valuation. Even though the claim may be disputed initially, such action will ensure that if and when it is established, interest on any sum to which the contractor is later found to be entitled will run from the earliest possible date. During the course of negotiations with the engineer, the contractor should of course provide any information or evidence of cost etc. that may be required. Having done so it is reasonable for him to expect that the engineer will allow or reject the claim without unnecessary delay, and where the claim is allowed, will certify the payment due. Where, as is often the case, the claim is ongoing – such as one resulting from a continuing delay – the contractor may reasonably expect interim payments to be certified to cover the costs incurred up to the date of each monthly valuation.

Where the engineer requires the contractor to provide details of the pricing of relevant work, such as where a variation requires the adjustment of a contract rate, the contractor should provide full pricing information. The practice among some contractors of resisting such requests on the ground that the information is confidential is unwise and unhelpful. The engineer requires the information solely for the purpose of evaluating the claim, and he is not entitled to use it for any other purpose, hence there is no loss of confidentiality. Furthermore, should the claim develop into a dispute referred to arbitration the employer's advocates will be entitled to demand production of relevant information.

## 11.6  *Adjudication*

Before the Housing Grants, Construction and Regeneration Act 1996 (the Construction Act) became law the verb 'to adjudicate' had a meaning defined in the dictionary as 'to determine judicially; to award'. The fourth and earlier editions of this book used the word in that sense. However the Construction Act has since given the word a special meaning in the construction industry by defining a procedure – called *adjudication* – for the rapid determination on an interim basis, by way of an adjudicator's *decision*, of disputes arising from construction contracts. Such decisions are enforceable immediately through the courts, notwithstanding any provision in the contract for arbitration.

Section 108 of the Construction Act requires construction contracts, as defined in the Act, to include a right of any party to the contract to refer disputes to adjudication in accordance with a procedure also defined in the Act. Where a construction contract does not include such provision, section 108(5) of the Act provides that the adjudication provisions of the Scheme for Construction Contracts (England and Wales) Regulations 1998 (SI 1998 No 649) (the Scheme), which came into force on 1 April 1998, apply to the contract. Both section 108 of the Act and the Scheme (see Appendix D) provide for adjudication under rules which provide that:

- notice of adjudication may be given by any party to the contract at any time;

- the adjudicator is appointed within seven days of the notice;
- the adjudicator's decision is to be given within 28 days of the notice, which period may be extended by up to 14 days with the consent of the party by whom the dispute was referred;
- the adjudicator's decision shall be binding on the parties, and they shall comply with it immediately upon delivery of the decision to the parties, until the dispute is finally determined by legal proceedings, by arbitration, or by agreement.

Prior to its enactment, the Construction Act, and in particular section 108 thereof, gave rise to strongly expressed misgivings by many eminent construction lawyers and arbitrators, partly on the ground that there had been insufficient effective consultation within the construction industry of such far-reaching and untried proposals. It was not until nearly two years after the Construction Act became law that section 108 was brought into operation, the delay having apparently been caused by the need to finalise the Scheme to which strong objections had been expressed.

In Volume II of *Contemporary Issues in Construction Law Construction Law Reform: A plea for sanity* edited by Professor John Uff QC and including contributions by Professor Ian Duncan Wallace QC, Donald Valentine, Robert Gaitskell QC, Ian Menzies, John Sims and other eminent lawyers and arbitrators, serious doubts were expressed as to the wisdom and likely efficacy of the adjudication proposals. In particular attention was drawn to:

- the inconsistency between the philosophy of the Arbitration Act 1996, which emphasised party autonomy, and that of the Construction Act, which imposed statutory intervention upon construction contracts;
- the likely outcome of the adjudication provisions in destroying the concept of the impartial engineer;
- duplication of existing dispute resolution machinery;
- generation of a greater number of formal disputes.
- the definition of a construction project which is in some cases related to the end use of the project rather than its construction.

Notwithstanding the reservations that had been expressed during the consultative stages the legislation implementing compulsory adjudication procedures became law on 1 April 1998. During the two and a half years that have elapsed since that enactment Professor Uff's predictions have in general proved to be accurate. Many contractors and subcontractors have made use of the adjudication procedure and have generally, where decisions in their favour have not been honoured, been successful in their appeals to the court for enforcement.

In *Macob Civil Engineering Ltd* v. *Morrison Construction Ltd* [1999] BLR 93 the main contractor, Morrison Construction Ltd (MCL), failed to comply with an adjudicator's decision awarding Macob an immediate payment of £300,000 plus VAT plus interest and fees. The subcontractor (Macob) sought enforcement of the decision in the Technology and Construction Court. MCL submitted that it was entitled to a stay of the court action pursuant to section 9 of the Arbitration Act 1996, that the adjudicator had been guilty of procedural errors in breach of the rules of natural justice and that summary judgment was inappropriate in the circumstances. Upholding the adjudicator's decision, Dyson J (now Dyson LJ) held that:

'The intention of Parliament in enacting the Act was plain. It was to introduce a speedy mechanism for settling disputes in construction contracts on a provisional interim basis, and requiring the decisions of adjudicators to be enforced pending the final determination of disputes by arbitration, litigation or agreement...'

Later in the judgment he stated:

'I do not consider that the mere fact that the decision may later be revised is a good reason for saying that summary judgment is inappropriate.'

In *Outwing Construction Ltd* v. *H. Randell & Son Ltd* [1999] BLR 156, Judge Humphrey Lloyd QC held that where the defendant main contractor had failed to pay in accordance with an adjudicator's decision until a few hours before the plaintiff subcontractor's application to the court for enforcement was heard, the plaintiff was justified in pursuing its application for costs.

In *Northern Developments (Cumbria) Limited* v. *J&J Nichol* [2000] BLR 158, a subcontractor J & J Nichol (JJ) having been denied payment by the main contractor (NDL) withdrew from the site and sought and obtained an adjudicator's decision in its favour. NDL contended that JJ's claim should be reduced by reason of defective work and of delays resulting from JJ's withdrawal from the site, which NDL contended to constitute repudiation of the contract. Upholding the adjudicator's decision Judge Bowsher held that it was 'clearly the intention of the Act to exclude the consideration of set-offs arising after the due date for the making of interim payment.'

He also held that the Construction Act gave the adjudicator no implied power to award costs, but that on the facts of the case both parties had, by each submitting a claim against the other for costs in the adjudication, impliedly agreed that the adjudicator had jurisdiction to award costs. Judge Bowsher stated

'The Scheme by paragraph 25 provides that the Adjudicator is entitled to payment of such reasonable amount as he may determine by way of fees and expenses. The same paragraph gives him the power to apportion liability for the payment of his fees by the parties. Nowhere in the Act or in the Scheme is the Adjudicator given power to order one party to the adjudication to pay the costs of the other.'

If the purpose of the adjudication provisions is to provide contractors, subcontractors and others involved in long-term construction contracts which have provision for interim payments to be authorised by a third party, with a procedure for appealing against unfair exercise of that third party's powers, then it appears from its initial reception to have achieved its intended purpose. Some commentators have aptly described the adjudicator's decision as being comparable with certificates of the engineer or the architect during the course of construction, in that both are intended to be enforced without delay, and neither purports to be final.

There is however an important difference between the engineer or architect on the one hand, and an adjudicator on the other. In the former appointment, the engineer or architect is expected to act impartially in dealing with contentious matters between the employer and the contractor, and indeed has a duty to do so defined by the House of Lords in *Sutcliffe* v. *Thackrah* (1974) 4 BLR 16 in which Lord Morris of Borth-y-Gest said of the architect:

'Being employed by and paid by the owner he unquestionably has in diverse ways to look after the interests of the owner. In doing so he must be fair and he must be honest. He is not employed by the owner to be unfair to the contractor.'

In the case of the adjudicator, however, the duty of impartiality is a statutory requirement of section 108(2)(e) of the Construction Act, while the Scheme requires, in regulation 4, that 'the adjudicator shall not be an employee of any of the parties to the dispute and shall declare any interest, financial or otherwise, in any matter relating to the dispute'.

It has been recognised for many decades that the decisions of an engineer or the architect employed and paid, in one way or another, by the employer, may not always be as impartial as the contractor is entitled to expect. Engineers and architects have the difficult task, when disputes arise, of representing their client, who is also the employer under the contract, and simultaneously of acting quasi-judicially in determining the matters in dispute. Generally engineers and architects cope with that dichotomy with remarkable skill; but there are inevitably occasions when the contractor believes, rightly or wrongly, that his treatment has been unfair. That belief may be even more likely to arise in the case of a dispute between a contractor and his subcontractor, where the engineer under the main contract may have little if any function or interest in the subcontract dispute. Clearly an adjudicator has an important role in such situations, and it is not surprising that contractors, and especially subcontractors, have not been slow in invoking the protection given them under the Construction Act.

An alternative procedure may sometimes be available to contractors and subcontractors who believe they have been treated unfairly by the engineer or architect, under section 39 of the 1996 Arbitration Act, under which the arbitrator may, subject to the agreement of both parties, be empowered to order provisional relief. Both the ICE Arbitration Procedure and CIMAR include agreements conferring such power on the arbitrator (see Chapter 2 subsections 2.8.12 and 2.8.13). Where neither of those procedures applies it is unrealistic to expect that the respondent will willingly agree, after a dispute has arisen, to confer a power upon the arbitrator which he knows will be used to his own detriment.

Furthermore it is reasonable to assume that the adjudication provisions of the Construction Act envisage that each dispute will be dealt with as it arises, providing a rapid interim settlement, which may or may not be challenged in arbitration or litigation. Conversely, arbitration envisages the collection together of all such disputes for reference to arbitration after the work has been completed: a procedure that has the merits of avoiding multiple arbitrations and the diversion of key staff from their prime duties during construction.

It remains to be seen whether or not the adjudication procedure in the Construction Act and in the Scheme will remain valid after 1 October 2000, when the Human Rights Act 1998 will become fully effective in the UK. Article 6(1) of the Convention of Human Rights provides that

> 'In the determination of his civil rights and obligations ... everyone is entitled to a fair and public hearing within a reasonable time by an independent and impartial tribunal established by law ...'

The adjudication procedure of the Scheme has already survived allegations that it breaches the rules of natural justice (see, for example the *Macob* case referred to above), but the wording of the Convention is slightly different from the common definition of 'natural justice' and could conceivably result in a major reversal of the established case law.

## 11.7 *Disputes*

A dispute is defined under clause 66 of the ICE Conditions as arising when one party serves on the other a 'notice of dispute', and some other forms of contract make similar provision. Usually it is unwise for the contractor to invoke clause 66 until after attempts to reach a negotiated settlement of the disputed issues, and further-more arbitration should not be invoked unless a party is confident that it has both a valid case and the evidence to substantiate it. Clause 66 itself provides for several attempts to reach a settlement, by the engineer, by conciliation or by adjudication. The procedures to be followed are outlined in Chapter 2, section 2.8, and care should be taken by both parties to follow those procedures to the letter. In particular the contractor should take careful note of the time limits prescribed under the clause, because failure to give notice of dis-satisfaction with the engineer's decision and written 'notice to refer' within the prescribed time limits will result in the engineer's deci-sion being deemed to have been accepted. When this happens, the contractor is left with no means of enforcing his claims, either in arbitration or in litigation.

Whether or not attempts are made to resolve the disputed matters by negotiation, it is often unwise for a contractor to allow arbitration procedures, once commenced, to be delayed by such negotiations, because settlements are often precipitated by the imminence of the arbitration hearing. Negotiations towards a settlement may con-tinue concurrently with the arbitration proceedings; and the various stages of the proceedings often lead to a settlement when each party becomes aware, through statements of case and through disclosure of documents, of the strength of the opposing party's case.

## 11.8   *Appointment of the arbitrator*

Either party may serve on the other a 'notice to concur' in the appointment of an arbitrator; and by so doing the parties have an opportunity to choose a person having the most suitable qualifications for that appointment (see Chapter 3, section 3.4). The ICE List of Arbitrators provides a valuable source of information to those seeking suitable nominees, and it is suggested that selection by the parties of a candidate from that list is generally to be preferred to appointment by the president of a professional body, because the parties are more fully aware of their own needs as to the technical qualifications and experience required of their arbitrator. Where the dispute arises from a building contract, detailed information as to the *curricula vitae* of architects or quantity surveyors may not be so readily available; but the Chartered Institute of Arbitrators publishes a Register of Arbitrators which contains similar information to that in the ICE list, in respect of arbitrators in general, covering persons of all basic professions who have qualified with the Chartered Institute. Selection of suitable nominees should, it is suggested, have regard not only to the type of construction work from which the dispute arose, but also to other factors, such as experience both as engineer, architect, or quantity surveyor, and as a director or a senior employee, such as an engineer or a quantity surveyor, of a contractor.

Having identified possible candidates for appointment as arbitrator, either by way of nominees by the parties themselves or by obtaining recommendations from a professional body, it may sometimes be advantageous for the parties to seek further details of each nominee's experience, either in writing or at an interview. Where such a procedure is adopted it is important that both parties should be involved. All written communications between the parties and the candidates should be made available to both parties; and where it is decided to interview candidates, both parties should be present in person or by representation. Where such procedures do not lead to the identification of a candidate acceptable to both parties, they can only resort to a presidential nomination.

## 11.9   *Choice of advocate*

Subject to any agreement the parties may have entered into, each party is free to appoint any person it wishes as its advocate, with the proviso that it must forewarn the opposing party of its intention. In major disputes – say those involving sums running to seven or more figures at 2000 prices – a party's main concern is that its advocate should be such as to be most likely to succeed, costs being a relatively minor factor. Hence it is usual in such cases for each party to appoint leading counsel from one of the chambers specialising in construction contract disputes: which counsel will usually require in addition the appointment of a junior who is responsible to his leader for most of the routine work. For smaller disputes junior counsel from specialist chambers may be the appropriate choice, while for disputes involving sums of five figures or less the choice is usually between a solicitor experienced in construction contract disputes or a technically qualified claims consultant, such as a quantity surveyor, again with suitable experience, or a member of the party's own staff. The appointment of solicitors or counsel without a specialised knowledge of construction contract disputes is usually unwise. Such contracts are highly complex and quite different from other forms of commercial contract: so that even a solicitor specialising in litigation is likely to find his knowledge inapplicable to the requirements of a construction contract dispute.

Apart from their knowledge of procedural matters, the main advantages to be gained by the appointment of specialist counsel are their skill in dealing with complex legal issues; in conducting cross-examination where issues both of fact and of opinion are in dispute, and in logical analysis of both the claim and the defence. It is frequently advantageous for a party to plead alternative cases, so that failure in its primary submission does not necessarily result in abandoning a claim. Equally important may be alternative defences: especially, for example, a challenge to the quantum of a claim as an alternative to a primary defence denying liability.

## 11.10 Offers to settle

Both parties should be aware of the significance of offers to settle the dispute, and of the benefits to be gained by making such offers, which may be made at any time after the dispute has arisen. In addition to the possibility that the offer may be accepted, thereby obviating further expenditure on lawyers and on the arbitrator, a party who makes an offer of such magnitude that it should, in the opinion of the arbitrator (when he reaches the stage of dealing with costs) have been accepted, will generally result in costs incurred subsequently to the date of the offer, or shortly thereafter, being borne by the offeree.

The initiative towards a negotiated settlement should usually be taken by the respondent, who should, as soon as it becomes apparent that the claims against him are valid and are likely to succeed in arbitration, make an offer to settle the claims. He should form his own assessment of the value of the claims, erring on the side of generosity in order to allow for the uncertainty of his succeeding in his defence and hence of the risk that he may be ordered to pay all costs in addition to the value of the claims if the arbitration proceeds. That sum, together with an offer to bear such reasonable costs as the claimant may at that stage have incurred, or an additional sum representing the respondent's assessment of such costs, should be offered to the claimant, in the form of a Part 36 offer.

Upon receipt of such an offer the claimant has three options: to accept it, to reject it, or to try to negotiate a higher figure. Before rejecting the offer out of hand the claimant should make his own assessment of the value of his claims, recognising that even though his claims may be valid, the burden of proof rests upon him, and that unless comprehensive records have been maintained he may have difficulty in establishing the validity and value of those claims. Unless the offer is so low as to be derisory the contractor is unwise to reject it without at least making a counter-offer to accept a larger sum, because the original offer does at least imply a willingness to negotiate a settlement. Furthermore the contractor should recognise that failure to accept an offer in the form of a Part 36 offer puts him at risk in respect of all costs incurred subsequently, which costs may become substantial if the arbitration proceeds to a hearing.

# APPENDICES

# APPENDIX A
# SPECIMEN DOCUMENTS

The specimen letters, orders, agreements, applications, notices, directions, statements, awards and other documents included in this appendix have been chosen in order to provide suggested formats for the wide range of documents sometimes needed in arbitration. Of these documents only a few are likely to be relevant to any particular arbitration; and where a document is so relevant, it is likely to require a substantial degree of editing in order to suit the arbitration in question. This is because of the inclusion in the specimens of matter that may only rarely be relevant to any given arbitration, in order to provide a reminder of such matter in the unlikely event that it may be needed.

It follows that in using the documents great care should be exercised firstly in selecting a format that is relevant to the particular need, and secondly in editing that format so as to exclude irrelevant matter and to revise it to suit the particular circumstances of the reference.

# SD/1
## *Ad hoc* arbitration agreement

*IN THE MATTER OF THE ARBITRATION ACT 1996*

*AND*

*IN THE MATTER OF AN ARBITRATION BETWEEN*

<u>C</u>

CLAIMANT

*AND*

<u>R</u>

RESPONDENT

## ARBITRATION AGREEMENT

We the undersigned HEREBY AGREE to refer to arbitration a dispute that has arisen from a contract between us dated     199    for the construction of     and we HEREBY APPOINT Isambard Kingdom Brunel, Chartered Civil Engineer, of 'The Homestead' Clifton, Bristol, to be Arbitrator in the reference.

Signed on behalf of the Claimant by:

Dated this      day of      2001

Signed on behalf of the Respondent by:

Dated this      day of      2001

# Application for stay of court proceedings

*IN THE HIGH COURT OF JUSTICE*
*QUEEN'S BENCH DIVISION*

2000 W No 1234

*BETWEEN*

Wright, Charlie & Company

*PLAINTIFF*

*AND*

Universal Finance Corporation PLC

*DEFENDANT*

## APPLICATION

1. This is an application under Part 11 of the Civil Procedure Rules 1998 by the defendant for a stay of the proceedings pursuant to Section 9 of the Arbitration Act 1996.

2. The defendant has filed an acknowledgement of service under Rule 11(3) of the CPR.

3. The matters in dispute arise from a contract under seal dated 15 February 1998; which contract includes, in Clause 66, an arbitration agreement as defined under section 6 of the Arbitration Act 1996.

Signed on behalf of the defendant by:

Dated this 20th day of January 2001

Copy served on plaintiff on 20 January 2001

*IN THE MATTER OF THE ARBITRATION ACT 1996*

*AND*

*IN THE MATTER OF AN ARBITRATION BETWEEN*

<u>C</u>

*CLAIMANT*

<u>AND</u>

<u>R</u>

*RESPONDENT*

## EXCLUSION AGREEMENT

We the undersigned having referred to arbitration a dispute that has arisen from a contract between us dated       199    for the construction of                hereby agree:

1.  that the power of the court pursuant to section 45 of the Arbitration Act 1996 to determine any question of law arising in the course of the proceedings, and

2.  that the power of the court pursuant to section 69 of the Arbitration Act 1996 to give leave to appeal to the court on a question of law arising from an award

be hereby excluded.

Signed on behalf of the Claimant by:

Dated this       day of       2001

Signed on behalf of the Respondent by:

Dated this       day of       2001

# Notice to refer and notice to concur

From:     The Universal Construction Company Ltd

To:       The Department of the Environment, Transport and the Regions

23 November 2000

Dear Sirs

**Contract for the construction of road bridge: Gosport to Ryde, IOW**

A matter of dissatisfaction with an act of the Engineer's Representative having arisen and having been referred to the Engineer pursuant to clause 66(2) of the contract (namely the ICE Conditions of Contract, Seventh Edition) by our letter dated 10 September 2000, we confirm having thereafter, by letter dated 9 October 2000 and upon receipt of the Engineer's decision, given Notice of Dispute pursuant to clause 66(3) of the contract.

We now give this our Notice to Refer the dispute to arbitration, pursuant to clause 66(9) of the contract.

We also give Notice to Concur, pursuant to clause 66(10) of the contract, in the appointment of one of the following gentlemen to be arbitrator in the reference, or to propose an alternative person for our consideration:

John Smeaton      590 Victoria Street, London

Thomas Telford    20 High Street Blanktown, Anglesea

John Rennie       40 Smith Street, Littletown, Wessex

Failing agreement being reached within one month of the date of this letter we shall apply to the President of the Institution of Civil Engineers to appoint an arbitrator.

Yours faithfully

# Application to President to appoint an arbitrator

From:    The Universal Construction Company Ltd   *(Claimant)*

To:    The President, The Institution of Civil Engineers

Copy to:    The Department of the Environment, Transport and the Regions
(Ref: XYZ)   *(Respondent)*

29 December 2000

Dear Sir

**Construction of Road Bridge: Gosport to Ryde, Isle of Wight**

A dispute has arisen from our contract with the Department of the Environment for the construction of the above bridge, on which work has reached the stage of substantial completion. The contract incorporates the ICE Conditions of Contract, Seventh Edition, which as you will know provides for the appointment of an arbitrator, failing agreement, by the President or a Vice-President of the Institution of Civil Engineers.

We have submitted to the Respondent, in our letter dated 23 November 2000 (copy enclosed) the names of three persons who would be acceptable to us for appointment as arbitrator. The Respondent has failed to reply to that letter.

We accordingly request that you appoint an arbitrator to determine the matters in dispute; which matters concern the evaluation of additional excavation and reinforced concrete in deepened foundations to the structure, together with the provision and driving of steel sheet piling needed to support the sides of the excavation; the costs incurred in delay to the contract works arising from the above and from numerous minor variations; and the evaluation of certain other varied works.

The sum in dispute amounts to £8.7m. We suggest that the arbitrator should be a civil engineer having experience in the construction of major prestressed concrete bridges over water; preferably both as an Engineer under the contract and as a Contractor.

Yours faithfully

From:      Robert Stephenson FRS FICE FCIArb

To:      The Universal Construction Company Limited

The Department of the Environment, Transport and the Regions

22 January 2001

Gentlemen

**Arbitration between the Universal Construction Company Limited (*Claimant*) - and – The Department of Transport (Respondent) : Road Bridge: Gosport – Ryde**

I am appointed by the President of the Institution of Civil Engineers to be Arbitrator in the above reference. I hereby accept the appointment.

Proposed terms of my appointment are set out in the Appendix hereto. Both parties are invited to indicate their agreement to the proposed terms by signing the duplicate of the Appendix and returning it to me. The terms will become effective only if both parties to the reference agree thereto. In the event that either or both parties fail so to agree within 28 days of the date of this letter I shall in due course determine the amount of my costs pursuant to section 63 of the Arbitration Act 1996; and meanwhile I shall require the Claimant to provide security for those costs pursuant to section 38 of the Arbitration Act 1996.

It is my intention to convene a preliminary meeting for directions during the week commencing 19 February 2001 at the Institution of Civil Engineers or other suitable venue in London, commencing at 11.00 am. Both parties are invited to advise me, no later than 28 January 2001, if any date during that week is unsuitable for the meeting.

When replying to this letter, and whenever writing to me in the future, the parties are required to send a copy of their letter to the other party, indicating to me that they have done so. Letters may, in cases of urgency, be sent to me by fax, provided that a copy is also sent by fax to the other party. I shall not use the telephone for communication with either party, and I shall not accept any telephone call from either party.

Yours faithfully

*Arbitrator*

Arbitration between the Universal Construction Company Limited (*Claimant*) – and – The Department of The Environment, Transport and the Regions (*Respondent*)

## FEES AND TERMS IN RESPECT OF APPOINTMENT AS ARBITRATOR

(1)  A lump sum *appointment fee* of £X plus VAT thereon payable at the commencement of the reference; plus

(2)  An *hourly rate* of £Y plus VAT thereon for each hour spent on or in connection with the arbitration, including but not limited to administration, travelling, and reading papers; plus

(3)  A *daily rate* of £Z plus VAT thereon for each day spent on or allocated to hearings, meetings, inspections, and site visits; plus

(4)  The cost of all travelling expenses and other outgoings incurred in the reference.

*Provided that:*

(a)  All accounts shall be paid within twentyone days of the invoice date. In the event of failure to pay by the due date the Arbitrator may treat the appointment as repudiated. Interest on overdue payments shall accrue at 2 per cent per annum over base rate.

(b)  Payments on account of charges may, at the Arbitrator's discretion, be invoiced at the commencement of the arbitration and thereafter at quarterly intervals, and may include items in respect of time allocated to meetings or hearings.

(c)  Should the arbitration continue for more than one year after the date of the appointment the hourly and daily rates referred to above may be varied in accordance with the General Index of Retail Prices (RPI).

(d)  The *daily rate* will be charged in respect of any date or dates fixed for a meeting or hearing, and subsequently vacated at the request of, or by default of, either or both parties. In such cases the charges for vacated periods will be reduced by the following percentages:

| | |
|---|---|
| Periods vacated more than 6 months in advance | Reduction 90% |
| Periods vacated more than 3 months in advance | Reduction 75% |
| Periods vacated more than 1 month in advance | Reduction 50% |
| Periods vacated less than 1 month in advance | Reduction 25% |

(e)  The appointment fee and any other charges payable before the date of the Award will be the responsibility of the Claimant unless the Respondent agrees otherwise. Such charges will in due course be borne by such party or parties as may be directed in an Award of costs, pursuant to Section 61 of the Arbitration Act 1996.

(f)  Any negotiated settlement shall be incorporated in an Agreed Award.

*The above fees and terms are hereby agreed on behalf of the Claimant/Respondent*

*Signed:*

Date:                    200...

# Agenda for preliminary meeting

Arbitration between  *(Claimant)* – and -  *(Respondent)*

### AGENDA FOR PRELIMINARY MEETING ON  2001

## 1. APPEARANCES:

|  | CLAIMANT | RESPONDENT |
|---|---|---|
| Name | | |
| Appointment | | |
| | | |
| Name | | |
| Appointment | | |

## 2. CONTRACT:

Form, edition etc.

Date executed

Arbitration clause therein

Copy for Arbitrator's records

## 3. OUTLINE OF DISPUTE:

Claim: subject

Approximate value

Counterclaim (if any): Subject

: Approximate value:

## 4. PROCEDURAL MATTERS:

S.34(2)(a)   Venue for meetings, hearings etc

S.34(2)(b)   Language of proceedings: translations where needed

S.34(2)(c)   Written Statements of Case

S.34(2)(d)   Disclosure of documents

S.34(2)(f)   Application of strict rules of evidence

# Agenda for preliminary meeting

| | |
|---|---|
| S.34(2)(g) | Arbitrator to ascertain facts and law? |
| S.34(2)(h) | Written or oral submissions? |
| S.35(1)(a) | Consolidation with other proceedings? |
| S.35(1)(b) | Concurrent hearings? |
| S.36 | Form of representation of parties? |
| S.37(1) | Appointment by arbitrator of experts/legal advisers/assessors? |
| S.38(3) | Security for costs to be provided by Claimant? |
| S.38(5) | Examination of witnesses on oath? |
| S.38(6) | Preservation of evidence? |
| S.39 | Provisional relief (power available only by parties' agreement)? |
| S.45 & 69 | Exclusion of power of court to determine questions of law? |
| S.46(1) | Choice of relevant substantive law? |
| S.46(2) | Determination of disputes on considerations other than law (*equity*)? |
| S.47 | Separate awards covering separate issues? [I shall exclude costs from the ambit of my first award, and shall make a separate award of costs thereafter having heard both parties on the question of award of costs.] |
| S.48 | Form of award(s): declaratory; monetary; performance; rectificatory? |
| S.49 | Award of simple or compound interest? |
| S.63(3) | Determination by arbitrator of amount of recoverable costs? |
| S.65(1) | Limitation of recoverable costs? |

## 5. PROGRAMME OF EVENTS:

Statement of Claim no later than                           2001

Statement of Defence (and Counterclaim?) within            weeks thereafter

Statement of Reply (and Defence to CC?) within             weeks thereafter

Reply to Defence to Counterclaim within                    weeks thereafter

Requests for Particulars within     weeks of Statement referred to

Particulars, where requested/ordered, within     weeks of request/order

## 6. DISCLOSURE OF DOCUMENTS: TIME ALLOCATED:

Lists of Documents exchanged within     weeks of final Statement of Case

Inspection on 7 days notice

# Agenda for preliminary meeting

Preparation of agreed bundles within      weeks thereafter

7. **PROOFS OF EVIDENCE: WITNESSES OF FACT:**

   Date for exchange of proofs of evidence (to stand as evidence-in-chief)      2001.

8. **EXPERT EVIDENCE**

   Appointments, dates for reports: meetings: agreed reports

9. **OPENING SUBMISSIONS IN WRITING:**

   Date(s) for submission: Claimant:                    Respondent:

10. **ARRANGEMENTS FOR HEARING:**

    Commencement date

    Number of working days reserved and dates

    Sitting hours

    Venue (to be reserved by Claimant)

    Inspection of real evidence on site?

    Transcript or tape recording of evidence/addresses?

11. **GENERAL DIRECTIONS:**

    Copies of letters to Arbitrator, by post or by fax, to be sent to other party and originals to be marked accordingly

    No communication with Arbitrator by telephone except in emergency

    Use fax, with copy to other party, for urgent communications

    Figures to be agreed as figures where possible

    Plans, photographs and documents to be agreed where possible

    Liberty to apply

    Costs of this meeting and Directions to be costs in the reference

12. **TERMS OF APPOINTMENT:**

    Agreed in writing by Claimant on:

    Agreed in writing by Respondent on:

*IN THE MATTER OF THE ARBITRATION ACT 1996*
*AND*
*IN THE MATTER OF AN ARBITRATION BETWEEN*

<u>C</u>

CLAIMANT

*AND*

<u>R</u>

RESPONDENT

## PROCEDURAL DIRECTIONS

Upon hearing learned counsel for both parties, and by consent, the following Directions are given:

1. There shall be an exchange of written Statements of Case as follows:
    1.1 Statement of Claim shall be delivered no later than        2001.
    1.2 Statement of Defence [and Counterclaim] shall be delivered within [       ] weeks thereafter.
    1.3 Statement of Reply (if any) [and Defence to Counterclaim] shall be delivered within [       ] weeks thereafter.
    [1.4 Statement of Reply to Defence to Counterclaim, if any, shall be delivered within [       ] weeks thereafter.]
    1.5 Requests for Particulars, if any, shall be delivered within [       ] weeks of delivery of the Statement to which they relate.
    1.6 Particulars, where requested or directed, shall be delivered within [       ] weeks of the request or order.
2. There shall be Disclosure of Documents as follows:
    2.1 Within [       ] weeks of issue of the final Statement of Case the Claimant and the Respondent shall each deliver to the other a List of Documents/List of Files which are or have been in their possession or power relating to the matters in issue in this arbitration.

# Procedural directions

2.2 Each party shall be entitled to inspect the other party's documents on giving 7 days notice of intention to do so, and shall identify any documents therein upon which the party intends to rely.

2.3 Bundles of documents upon which the parties intend to rely shall be prepared jointly by the parties within [    ] weeks after identification of relevant documents.

3. Proofs of evidence of witnesses of fact shall be exchanged no later than    weeks before the commencement of the Hearing and shall stand as evidence-in-chief of such witnesses.

4. Experts may be appointed by the parties provided that:

   4.1 No more than [    ] Experts may be appointed by either party, of whom one shall be an expert in quantum, and one an expert in  .

   4.2 Experts of like disciplines shall meet *without prejudice* in order to agree facts and technical issues so far as may be possible; and shall exchange Reports no later than [    ] 2001.

   4.3 Where possible Experts of like disciplines shall prepare a joint Report on matters on which they are agreed, no later than    2001. Matters which are not agreed may form the subject of separate individual Reports.

   4.4 Experts' Reports shall stand as evidence-in-chief of the Expert.

5. The Claimant's opening address shall be submitted in writing no later than    2001. The Respondent's opening address shall be submitted in writing no later than [    ] 2001.

6. Arrangements for the Hearing are confirmed as follows:

   6.1 Commencement date: [    200... ]

   6.2 Number of days reserved and dates:  .

   6.3 Sitting hours:

   6.4 Venue: The Institution of Civil Engineers, Great George Street, London SW1P 3AA or such other venue as may be agreed. Venue to be reserved by Claimant.

   6.5 An inspection of real evidence at the site of the works, at which both parties shall be represented, will take place on [    ] 2001, commencing at [    ] hrs.

   6.6 Both parties have given notice of their intention to be represented by junior/leading counsel. Any change of either party's intention shall be notified to me and to the other party in sufficient time to allow for any resultant change that other party may wish to make in its own representation.

   6.7 A transcript of the Hearing shall be made and shall be arranged by the Claimant in consultation with the Respondent.

7. I shall make a monetary Award save as to costs on all substantive matters in dispute (including interest on any sums that I may award) and shall thereafter provide an opportunity for the parties to address me, either orally or, if both parties so agree, in writing, before I make my Final Award as to costs.

8. The Claimant having so agreed I shall in due course exercise my power under section 63(3) of the Arbitration Act 1996 to determine the recoverable costs in this arbitration, where such costs are not agreed.

9. Figures shall be agreed as figures where possible.

10. Plans, photographs and documents shall be agreed where possible.

11. The parties when writing to me shall send a copy of their letter and of any enclosures thereto to the other party, indicating to me that they have done so.

12. There shall be no communication by telephone between either party and myself. Urgent matters shall be communicated to me and to the other party by fax.

13. I shall expect to receive copies of Statements of Case, Lists of Documents on which the parties intend to rely, and of other communications referred to in this Order, but not of routine correspondence between the parties. Both parties have liberty to apply in respect of the above or of any other matter.

14. The costs of this Order shall be costs in the reference.

Dated this [            ] 2001.

*John Smeaton*
*ARBITRATOR*

IN THE MATTER OF THE ARBITRATION ACT 1996

AND
IN THE MATTER OF AN ARBITRATION BETWEEN

Longspan Bridgebuilders PLC

*CLAIMANT*

*AND*

Universal Finance Corporation PLC

*RESPONDENT*

# STATEMENT OF CLAIM

1. The Claimant is a civil engineering contractor based in Wessex. The Respondent is a banker providing finance to road authorities and other organisations for major engineering and building projects.
2. By a contract under seal dated 15 February 1998 ('the Contract') the Claimant undertook to construct and complete certain works of civil engineering construction, namely a bridge over the river Avon, in consideration for payment by the Respondent of the Contract Price at the time and in the manner prescribed by the Contract.
3. The Contract incorporates the ICE Conditions of Contract, Sixth Edition, dated January 1991 ('the ICE Conditions'); under which the Claimant is the Contractor and the Respondent is the Employer.
4. Clause 12 of the ICE Conditions provides that the Contractor shall, subject to certain conditions, be paid the amount of any additional costs reasonably incurred by the Contractor by reason of encountering physical conditions or artificial obstructions such as could not reasonably have been foreseen by an experienced contractor: together with a reasonable percentage addition thereto in respect of profit.
5. Clause 60(7) of the ICE Conditions provides that in the event of failure by the Engineer to certify or the Employer to pay sums to which the Contractor is entitled, in accordance with Clause 60(2) of the ICE Conditions, the Employer shall pay to the Contractor interest compounded monthly for each day on which

any payment is overdue or which should have been certified and paid at a rate equivalent to 2% per annum above the base lending rate of the bank specified in the Form of Tender.

6.  During the progress of the works the Claimant encountered physical conditions, namely the presence of hard rock in ground required under the Contract to be excavated, for which no provision had been made under the Contract, and which could not reasonably have been foreseen by an experienced contractor.

7.  As a consequence of the said physical conditions the Claimant incurred additional costs in having to bring to the site and to employ equipment and labour for breaking out the rock; and in addition the Claimant incurred delay to the progress of the works, which resulted in further additional costs.

8.  The Engineer failed to certify payment to the Claimant of sums to which it is entitled pursuant to the said Clause 12; which sums amount to £4,456,664.

### PARTICULARS

**Additional Works and Delays**

| | |
|---|---:|
| Plant costs | 1,780,000 |
| Labour costs | 1,260,000 |
| Material costs | 790,000 |
| Site overheads: 18 weeks at £20,800/week | 374,400 |
| | £4,204,400 |
| Off-site overheads and profit at 6% of total | 252,264 |
| TOTAL | £4,456,664 |

9.  Accordingly the Claimant claims £4,456,664 under clause 12 of the contract, plus interest thereon at 2% above base rate, compounded monthly, pursuant to clause 60(7) of the Contract.

10.  I certify on behalf of the Claimant that the facts stated in this Statement of Claim are true to the best of my knowledge and belief.

Perry Mason

Served this 9th day of January 2001 by Jones & Smith of 560 Fleet Street London EC4, Solicitors for the Claimant

*IN THE MATTER OF THE ARBITRATION ACT 1996*

*AND*

*IN THE MATTER OF AN ARBITRATION BETWEEN*

Longspan Bridgebuilders PLC

*CLAIMANT*

*AND*

Universal Finance Corporation PLC

*RESPONDENT*

---

## STATEMENT OF DEFENCE

---

1. Paragraphs 1 and 2 of the Statement of Claim are admitted.

2. Save that the Contract also incorporates certain additional and special clauses, together with a Specification, a Bill of Quantities, a Site Investigation Report, and certain drawings, upon which the Respondent will rely for their full content and meaning, paragraph 3 of the Statement of Claim is admitted.

3. Save that the Respondent will rely upon the full content and meaning of the said Clause 12 of the ICE Conditions, paragraph 4 of the Statement of Claim is admitted.

4. Save that the Respondent will rely upon the full content and meaning of the said Clause 60(7) of the ICE Conditions, paragraph 5 of the Statement of Claim is admitted.

5. In respect of paragraph 6 of the Statement of Claim it is admitted that during the progress of the works the Claimant encountered physical conditions, namely the presence of hard rock in ground to be excavated; but it is denied that the said physical conditions could not reasonably have been foreseen by an experienced contractor. The Claimant could and should have foreseen the said conditions; and by failing to do so the Claimant failed to exercise the skill reasonably to be expected of an experienced contractor.

6. Paragraph 7 of the Statement of Claim is not admitted.

7. In respect of paragraph 8 of the Statement of Claim it is denied that the Engineer failed to certify, or that the Employer failed to pay, the sum of £4,456,664 or any other sum to which the Claimant is entitled.

8. Save as herein expressly admitted the Respondent denies each and every allegation contained in the Statement of Claim as if the same were expressly set out and specifically denied seriatim.

9. I certify on behalf of the Respondent that the facts stated in this Statement of Defence are true to the best of my knowledge and belief.

James Kavanagh QC

Served this sixth day of March 2001 by Newpastures & Co of 180 Plastic Buildings, Temple, London EC4, Solicitors for the Respondent.

*IN THE MATTER OF THE ARBITRATION ACT 1996*

*AND*

*IN THE MATTER OF AN ARBITRATION BETWEEN*

Longspan Bridgebuilders PLC

*CLAIMANT*

*AND*

Universal Finance Corporation PLC

*RESPONDENT*

## STATEMENT OF REPLY

1.  Save insofar as the Statement of Defence consists of admissions the Claimant joins issue with the Respondent in each and every allegation contained therein.

2.  It is denied that the Claimant could and should have foreseen the physical conditions encountered during excavation; and it is denied that the Claimant failed to exercise the skill reasonably to be expected of an experienced contractor, as alleged in paragraph 5 of the Statement of Defence.

3.  I certify on behalf of the Claimant that the facts stated in this Statement of Reply are true to the best of my knowledge and belief.

Perry Mason

Served this 27th day of March 2001 by Jones & Smith of 560 Fleet Street London EC4, Solicitors for the Claimant.

# SD/13
## Scott Schedule

| Item No | CLAIM | AMOUNT | DEFENCE | OFFER | ARBITRATOR'S FINDING |
|---------|-------|--------|---------|-------|---------------------|
| 1 | Extension of time: inclement weather | 6 weeks | Part of time lost foreseeable | 2 weeks | |
| 2 | Extension of time: late issue of drawings | 8 weeks | No delay caused | Nil | |
| 3 | Payment for ditto | £6,000 | No liability | Nil | |
| 4 | Variations: founds: extra time | 8 weeks | Admitted | 8 weeks | |
| 5 | Ditto: additional excavation | £12,000 | Quantity agreed. Rate reduced | £9,000 | |
| 6 | Ditto: alterations to steel sheet piling | £14,000 | Piling to increased depth should have been allowed for | Nil | |
| 7 | Variations: general delay and disruption | 20 weeks | Denied | Nil | |
| 8 | Ditto: additional labour costs | £16,000 | Denied | Nil | |
| 9 | Ditto: additional site oncosts | £4,000 | Denied | Nil | |
| | Etc. | | | | |

# Directions letter: small claim

From:     I Solomon FRICS FCIArb

To:       A Homeowner Esq

          S M Builder Esq

21 December 2001

Gentlemen

**Arbitration between Alan Homeowner (*Claimant*) – and – Sean M Builder (*Respondent*)**

I am appointed by a Vice-President of the Chartered Institute of Arbitrators ('CIArb') to be Arbitrator in the above reference. I hereby accept the appointment.

My charges will be at the rate of £X per hour for time during which I engage myself upon, or which I allocate to, the duties of the reference, plus the amount of all expenses incurred in the execution of those duties, plus Value Added Tax on the amount of my fees and expenses.

In connection with the appointment I have received copies of the following:

(a)   CIArb's *Application for the Appointment of an Arbitrator*, incorporating a *Schedule of Claim* completed and signed by the Claimant on 13 November 2001.

(b)   Letter of appointment and copy of CIArb letter dated 12 December 2001 notifying both parties of my appointment to be Arbitrator in this reference.

Having regard to the small amount of the claim in this arbitration I suggest that the parties adopt, by agreement, the following rules of procedure with a view to minimising costs:

1.   That neither party be represented by a lawyer or by an expert: and that each party pay its own costs in the arbitration, if any.

2.   That the matters in dispute be determined upon my consideration of written submissions and written evidence adduced by the parties, followed by a meeting with the parties and an inspection by myself, in the presence of both parties, of the matters giving rise to the dispute.

3. That at such inspection the parties be permitted to draw my attention to factual matters they wish me to note, but not to make any further representation. I would however have the right to ask either party questions where needed in order to clarify their written submissions and to ascertain relevant facts.

4. That the parties comply with the following programme, subject to the right of either party to apply to me for an extension of time on reasonable grounds; the amount of such extension being specified:

   4.1 Within 14 days of the adoption of this procedure the Claimant shall submit to me, with a copy to the Respondent, a Statement of Claim together with documentary evidence in support of the claim and any other submissions the Claimant may wish to make.
   Such Statement may refer to the documents already submitted as referred to above, must clarify to what extent any quotations submitted are alternatives to one another, and must state in respect of each item claimed:

      4.1.1 The defect forming the subject of the item

      4.1.2 The proposed remedy to the defect, and

      4.1.3 The cost of the remedy proposed. Such costs should generally be the lower or lowest of quotations submitted in evidence.

   Items may also be included in respect of any professional advice taken or any other expenses necessarily incurred in determining the nature and extent of the alleged defects and of the remedial work required. The total sum claimed must be stated.

   4.2 Within 14 days of receipt of the Claimant's submissions the Respondent shall submit to me, with a copy to the Claimant, a Statement of Defence together with documentary evidence in support of the Defence and any other submissions the Respondent may wish to make. Such Statement may challenge the validity of any item of the Claim, or its amount, or both validity and amount; and it may include other statements or allegations.

   4.3 Within 7 days of receipt of the Respondent's submissions the Claimant may submit to me, with a copy to the Respondent, any Reply he may wish to make to the Respondent's submissions. Any such Reply, which may be supported by evidence, must deal only with matters in the Statement of Defence, and must not include any fresh claims or allegations.

4.4   Thereafter I shall convene a meeting with both parties at the site of the works and on a date to be arranged, for the purpose of inspecting the subject-matter of the dispute and of my asking the parties such questions as may be necessary in order to ascertain relevant facts.

Both parties are required to advise me in writing within 14 days of receipt of this letter whether or not they agree to adopt the procedure set out in paragraphs 1 to 4 above. If both parties do so agree the procedure will be adopted.

If only one party does so agree, and if at that stage it appears appropriate that I should do so, I shall exercise my power pursuant to section 34 of the Arbitration Act 1996 to determine that the above procedure shall apply to this arbitration, with the exception of paragraph 1 above, for which the agreement of both parties is necessary.

If neither party agrees to the procedure proposed above I shall convene a preliminary meeting in London in order to ascertain the parties' wishes as to procedure. The costs of any such meeting shall be costs in the arbitration.

I do not intend to have any communication with either party without the knowledge of the other party. Accordingly I shall address all of my letters to both parties.

*When writing to me, the parties must send a copy of their letter and of any enclosures thereto to the other party, indicating on their letter to me that such copy has been sent.*

For similar reasons I shall not accept any telephone call from either party, and I shall not meet either party except in the presence of the other party. Urgent communications may be sent to me by fax, provided that a copy of the communication is sent simultaneously to the other party, by fax if possible.

Yours faithfully

*Arbitrator*

*IN THE MATTER OF THE ARBITRATION ACT 1996*

*AND*

*IN THE MATTER OF AN ARBITRATION BETWEEN*

Longspan Bridgebuilders PLC

*CLAIMANT*

*AND*

Universal Finance Corporation PLC

*RESPONDENT*

## PROOF OF EVIDENCE

**JOSEPH BLOGGS of 23 Railway Cuttings, East Cheam, will say:**

I have been employed by Longspan Bridgebuilders since 1980. I joined the company as a carpenter and worked on several building projects in London. In 1988 I was promoted to foreman carpenter and in 1994 I was again promoted, to general foreman. I worked in that capacity on the Avon Bridge site in Wessex from the commencement of work in June 1998 until completion in September 2000.

On 15 August 1998 excavation for the foundations of the west abutment was in progress, using a Hymac hydraulic digger which was loading material into lorries for disposal off site. The subsoil was mainly sandy clay, which was being excavated without difficulty and in accordance with the construction programme. However at about 11.00 am on that day the digger driver reported that he had encountered hard obstructions which the digger was unable to remove. I immediately contacted the agent, Mr Brassey, and we went together to the west abutment site to inspect the problem.

We found that the obstruction consisted of hard rock at about 4 metres above formation level, and it was clear that compressors and breakers would be needed to break up the rock. As the drawings indicated that all excavation would be in soft ground we had no rock-breaking equipment on site, and I immediately ordered the necessary equipment from the plant yard. It arrived on site at 4.00 pm on the same day.

It soon became clear that because of the rock excavation for the west abutment would take much longer than had been planned. I told the digger driver to excavate soft material where he could, exposing the rock, and then to move on to the foundations of pier no 1. However at that pier excavation rock was again encountered at a level some 2 metres above formation level, and the compressor and breakers had to be used to break it out. Additional plant was brought onto the site on 22 August 1998 in order to expedite the work.

Excavations for the foundations of the west abutment and pier no 1 were not completed until 19 January 1999. The construction programme indicated that the excavation of the west abutment was due to be completed on 8 September 1998, and of pier no 1 on 15 September 1998.

I certify that the facts stated in this witness statement are true to the best of my knowledge and belief.

Dated this twentieth day of March 2001

(Signed)    Joseph Bloggs
              General Foreman

*IN THE MATTER OF THE ARBITRATION ACT 1996*
*AND*
*IN THE MATTER OF AN ARBITRATION BETWEEN*

Longspan Bridgebuilders PLC

*CLAIMANT*

*AND*

Universal Finance Corporation PLC

*RESPONDENT*

---

## AGREED AWARD

---

*WHEREAS:*

1.1 By a contract under seal dated 15 February 1998 (the Contract) the Claimant undertook to construct and complete certain works of civil engineering construction; namely a bridge over the river Avon; in consideration for which the Respondent undertook to pay to the Claimant the Contract Price at the times and in the manner prescribed by the Contract.

1.2 The said Contract provides that any dispute between the parties that might arise from it should be referred to the arbitration of a person to be agreed upon by the parties; or, failing such agreement, to be appointed on the application of either party by the President of the Institution of Civil Engineers.

1.3 A dispute having arisen and following upon an application by the Claimant on 11 September 2000 the President of the said Institution did on 9 October 2000 appoint me, John Smeaton, Chartered Civil Engineer and Chartered Arbitrator, to be Arbitrator in the reference: which appointment I accepted by notice in writing to both parties on 16 October 2000.

# Agreed award

1.4 Under cover of the said letter dated 16 October 2000 to both parties I set down proposed terms of my appointment; which terms were agreed in the Claimant's letter dated 23 October 2000, and in the Respondent's letter dated 14 November 2000.

1.5 Following upon a Preliminary Meeting and the issue by myself of certain Procedural Directions the Claimant notified me, by letter dated 10 January 2001, that the parties had agreed upon terms of settlement of the matters in dispute; under which terms the Respondent would pay to the Claimant the sum of £2.76m plus the Claimant's costs: such costs if not agreed to be determined by the Arbitrator, and would pay the amount of my costs: in full and final settlement of all claims and counterclaims arising from the Contract.

1.6 By letter dated 12 January 2001 the Respondent confirmed its agreement to the matters set out in paragraph 1.5 above.

*NOW I THE SAID JOHN SMEATON DO BY AGREEMENT HEREBY MAKE AND PUBLISH THIS MY AWARD.*

*I AWARD AND DIRECT THAT:*

2.1 The Respondent shall pay to the Claimant the sum of £2,760,000 (Two million seven hundred and sixty thousand pounds) in full and final settlement of all claims and counterclaims referred to me herein.

2.2 The Respondent shall pay the Claimant's costs of the arbitration: such costs, if not agreed, to be determined by me.

2.3 The Respondent shall pay to me the amount of my fees and expenses; which fees and expenses I hereby determine in the sum of £2,800 plus Value Added Tax thereon of £490 making a total of £3,290 (three thousand two hundred and ninety pounds).

*The seat of this arbitration is England*
*Dated this 19th day of January 2001*

(signed)

John Smeaton
*Arbitrator*

*IN THE MATTER OF THE ARBITRATION ACT 1996*
*AND*
*IN THE MATTER OF AN ARBITRATION BETWEEN*

Longspan Bridgebuilders PLC

*CLAIMANT*

*AND*

Universal Finance Corporation PLC

*RESPONDENT*

# AWARD SAVE AS TO COSTS

*WHEREAS:*

1.01 By a contract under seal dated 15 February 1998 (the Contract) the Claimant undertook to construct and complete certain works of civil engineering construction; namely a bridge over the river Avon in the county of Wessex; in consideration for which the Respondent undertook to pay to the Claimant the Contract Price at the times and in the manner prescribed by the Contract.

1.02 The Contract provides that any dispute between the parties that might arise from it should be referred to the arbitration of a person to be agreed upon by the parties; or, failing such agreement, to be appointed on the application of either party by the President of the Institution of Civil Engineers.

1.03 A dispute having arisen the parties agreed to appoint me, John Smeaton, Chartered Civil Engineer and Chartered Arbitrator, to be Arbitrator in the reference: of which agreement the Claimant notified me on 11 August 2000 and the Respondent confirmed its agreement on 18 August 2000. I accepted the appointment by letter dated 21 August 2000 addressed to both parties.

1.04 Under cover of the said letter dated 21 August 2000 I issued proposed terms of my appointment; which terms were agreed in the Claimant's letter dated 28 August 2000, and in the Respondent's letter dated 15 September 2000.

1.05 A preliminary meeting was convened by me and took place at the Institution of Civil Engineers, London, on 16 October 2000, following upon which I issued, on 18 October 2000, certain Procedural Directions.

1.06 Following upon an exchange of Statements of Case and disclosure of documents in accordance with the said Procedural Directions a hearing was held at the Institution of Civil Engineers, London, commencing on 12 March 2001 and continuing until 23 March 2001: at which hearing the Claimant was represented by Mr Perry Mason of Counsel and the Respondent by Mr James Kavanagh QC.

*NOW I THE SAID JOHN SMEATON having heard and considered the evidence both oral and written adduced by both parties, and having heard and considered addresses by learned Counsel for both parties DO HEREBY MAKE AND PUBLISH THIS MY AWARD SAVE AS TO COSTS.*

*I FIND THE FOLLOWING FACTS:*

2.01 The Contract is based on the ICE Conditions of Contract, Sixth Edition, dated January 1991, (the ICE Conditions) without material amendment.

2.02 Clause 12(1) of the ICE Conditions provides that:

*If during the execution of the Works the Contractor shall encounter physical conditions ... or artificial obstructions which conditions or obstructions could not in his opinion reasonably have been foreseen by an experienced contractor the Contractor shall as early as practicable give written notice thereof to the Engineer.*

2.03 Clause 12(6) of the ICE Conditions provides that:

*... the Engineer shall if in his opinion such conditions or obstructions could not reasonably have been foreseen by an experienced contractor determine the amount of any costs which may reasonably have been incurred by the Contractor by reason of such conditions or obstructions together with a reasonable percentage addition thereto in respect of profit and any extension of time to which the Contractor may be entitled ...*

2.04   During the excavations for the west abutment and for pier no. 1 of the Avon Bridge the Claimant encountered hard rock which required the use of special equipment in order that the excavation could proceed; which equipment was provided by the Claimant without undue delay.

2.05   The Claimant incurred additional costs in providing and operating the said equipment, and suffered delay to the programme of work as a result of the presence of rock.

2.06   The Claimant gave notices to the Engineer in accordance with the requirements of Clauses 12 and 52(4) of the ICE Conditions and in due course submitted its claim in the sum of £4,456,664 plus interest thereon.

2.07   The Engineer rejected the Claimant's claim on the ground set out in Clause 12(5) of the ICE Conditions, which provides that:

> If the Engineer shall decide that the physical conditions or artificial obstructions could in whole or in part have been reasonably foreseen by an experienced contractor he shall so inform the Contractor ...

2.08   Following upon a formal request by the Claimant for an Engineer's decision under Clause 66 of the ICE Conditions the Engineer notified the Claimant of his rejection of the claim by letter dated 16 May 2000, in which he stated that documents available at the time of tendering gave a clear indication that rock was to be expected in the deep excavations.

2.09   The sole issue between the parties is the question whether or not the presence of hard rock in the ground in which the west abutment and pier no. 1 were to be founded could reasonably have been foreseen by an experienced contractor.

2.10   Mr Mason for the Claimant has drawn to my attention the Site Investigation Report made available to tenderers, in which the only mention of rock is at levels no higher than 58.8m above ordnance datum (AOD): and he points out that the lowest excavation level shown on the drawings is 61.5m AOD. Learned counsel submits that the Claimant was entitled to infer, from the information provided by the Engineer, that there would be no hard excavation.

2.11   Furthermore Mr Mason submits that the Civil Engineering Standard Method of Measurement applicable to the Contract requires that rock excavation be measured separately from excavation in soft ground and that no such separation had been provided for in the contract documents.

# Award save as to costs

2.12 Mr Kavanagh submits on behalf of the Respondent that under Clause 11(2) of the ICE Conditions:

> The Contractor shall be deemed to have inspected and examined the Site and its surroundings and information available in connection therewith and to have satisfied himself so far as is practicable and reasonable before submitting his Tender as to (a) the form and nature thereof including the ground and the sub-soil (b) the extent and nature of work ... necessary for constructing and completing the Works.

2.13 Learned counsel submits that the Claimant could and should have ascertained the nature of the ground likely to be encountered in the excavations; and that had the Claimant done so and made proper provision in its programme of works and by having the necessary plant on site the delay would have been avoided.

2.14 I am not persuaded by Mr Kavanagh's submissions. I have no doubt that an adequate site investigation would have disclosed the presence of rock at the locations and levels at which it was encountered.

2.15 Under Clause 11(3) of the ICE Conditions the primary responsibility for carrying out an adequate site investigation rests upon the Employer; as is made clear by the words:

> The Contractor shall be deemed to have ...based his tender on the information made available by the Employer and on his own inspection and examination all as aforementioned ...

2.16 It follows that the Contractor's responsibility is limited to the inspection and examination of the site: the Contractor is not required to investigate the subsoil.

2.17 Accordingly I find in favour of the Claimant on the sole issue between the parties.

2.18 The parties' experts on quantum have helpfully agreed upon the evaluation of the claim, should I find it to be valid, in the sum of £3,866,000.

2.19 In addition the parties' experts have agreed that the interest payable on that sum pursuant to Clause 60(7) of the ICE Conditions, up to and including 31 May 2001, amounts to £764,580; continuing thereafter at £1,015 per day.

2.20 As at the date of this award, namely 24 June 2001, the additional interest to which the Claimant is entitled amounts to £24,360: and accordingly the sum payable to the Claimant by the Respondent as at the date of this award is £4,654,940.

# Award save as to costs

2.21  After the date of this award interest shall accrue at 8 per cent per annum on the sum of £4,654,940; such interest amounting to £1,020 per day.

*I HOLD THAT:*

3.01  The Respondent is liable to the Claimant in the sum of £4,654,940 in respect of the principal sum and interest thereon as at the date of this award, namely 24 June 2001: plus interest at £1,020 per day thereafter until the date of payment.

*AND ACCORDINGLY I AWARD AND DIRECT THAT:*

4.01  The Respondent shall pay to the Claimant the sum of £4,654,940 (four million six hundred and fifty-four thousand nine hundred and forty pounds) plus £1,020 (one thousand and twenty pounds) for every day that shall elapse from the date of this award until the date of payment: and in addition the Claimant shall pay to the Respondent Value Added Tax on the sums so paid at the rate prescribed by statute, in full and final settlement of all claims referred to me herein.

4.02  The Parties' costs in the arbitration shall be paid by such party or parties as I may direct in a future award after having heard both parties on that issue.

4.03  The amount of my fees and expenses in this arbitration; which fees and expenses I hereby determine in the sum of £18,400 plus Value Added Tax thereon of £3,220 making a total of £21,620 (Twenty-one thousand six hundred and twenty pounds) shall be paid by such party or parties as I may direct after having heard both parties on that issue.

*The seat of this arbitration is England*

Dated this 24th day of June 2001

(signed)

John Smeaton
*Arbitrator*

_IN THE MATTER OF THE ARBITRATION ACT 1996_
_AND_
_IN THE MATTER OF AN ARBITRATION BETWEEN_

Longspan Bridgebuilders PLC

_CLAIMANT_

_AND_

Universal Finance Corporation PLC

_RESPONDENT_

# FINAL AWARD

_WHEREAS:_

1.01 By a contract under seal dated 15 February 1998 ('the Contract') the Claimant undertook to construct and complete certain works of civil engineering construction; namely a bridge over the river Avon in the county of Wessex; in consideration for which the Respondent undertook to pay to the Claimant the Contract Price at the times and in the manner prescribed by the Contract.

1.02 The Contract provides that any dispute between the parties that might arise from it should be referred to the arbitration of a person to be agreed upon by the parties; or, failing such agreement, to be appointed on the application of either party by the President of the Institution of Civil Engineers.

1.03 A dispute having arisen the parties agreed to appoint me, John Smeaton, Chartered Civil Engineer and Chartered Arbitrator, to be Arbitrator in the reference: of which agreement the Claimant notified me on 11 August 2000 and the respondent confirmed its agreement on 18 August 2000. I accepted the appointment by letter dated 21 August 2000 addressed to both parties.

1.04 Under cover of the said letter dated 21 August 2000 I issued proposed terms of my appointment; which terms were agreed in the Claimant's letter dated 28 August 2000, and in the Respondent's letter dated 15 September 2000.

1.05 Following upon interlocutory proceedings and a hearing I made an Award on 24 June 2001, in which I determined all of the issues in the arbitration save that of my award of costs.

1.06 Both parties agreed that I should make this my Final Award as to costs without the need for a further hearing; taking account only of the parties' written submissions.

1.07 Such submissions comprise an application dated 4 July 2001 by the Claimant for its costs; and a submission dated 18 July 2001 by the Respondent that I should, in making my award of costs, take account of the fact that the claim has not succeeded in its entirety.

*NOW I THE SAID JOHN SMEATON having read and considered the parties' written submissions DO HEREBY MAKE AND PUBLISH THIS MY FINAL AWARD. I FIND THE FOLLOWING FACTS:*

2.01 The Claimant has substantially succeeded in its claim.

2.02 The Respondent has not brought to my notice any offer made to settle the claim against it, or any other valid reason, why I should not award the Claimant its costs.

## I HOLD THAT

3.01 The fact that the Claimant did not recover 100% of its claim does not constitute a valid reason for deviating from the rule costs *follow the event*. In reaching this conclusion I have had regard to the judgment of Phillips J (as he then was) in *Channel Islands Ferries Ltd* v. *Cenargo Navigation Ltd (The* Rozel) (1994) in which the learned judge made it clear that only partial success in a claim did not constitute grounds for depriving a claimant of the entirety of its costs.

## AND ACCORDINGLY I AWARD AND DIRECT THAT:

4.01 The Respondent shall pay the Claimant's costs in the arbitration: such costs if not agreed to be determined by me.

4.02 The amount of my fees and expenses in this arbitration; which amount I have already determined in the sum of £18,400 plus Value Added Tax thereon of £3,220 making a total of £21,620, having already been paid to me by the Claimant, shall be reimbursed to the Claimant by the Respondent.

The seat of this arbitration is England
Dated this twentyfourth day of July 2001

(signed)

John Smeaton
*Arbitrator*

# Notification of award

From:       John Smeaton                              (*Arbitrator*)

To:         Longspan Bridgebuilders PLC              (*Claimant*)

And:        Universal Finance Corporation PLC         (*Respondent*)

24 June 2001

Gentlemen

**Arbitration between Longspan Bridgebuilders PLC (*Claimant*) – and – Universal Finance Corporation PLC (*Respondent*)**

I have made my award save as to costs in this reference and it is available for collection by or dispatch to the parties upon payment of my charges, which amount to £21,620 including Value Added Tax.

Provision will be made in a future award for reimbursement of the above charges by the party or parties found to be responsible for them in the event that the other party pays the charges in taking up this award.

Yours faithfully

*Arbitrator*

From:     Universal Finance Corporation PLC     *(Respondent)*

To:     Longspan Bridgebuilders PLC     *(Claimant)*

13 March 2001

*WITHOUT PREJUDICE SAVE AS TO COSTS*

Dear Sirs

**Arbitration between Longspan Bridgebuilders PLC** *(Claimant)* –
**and – Universal Finance Corporation** *(Respondent)*

Further to our informal discussions at our offices yesterday, at which your company was represented by Mr Brassey and Mr Collins, and UFC by Messrs Jones and Robertson, I write to confirm our offer in accordance with Part 36 of the Civil Procedure Rules 1998 to settle the above dispute.

While my company remains convinced that its defence to your claims is valid and likely to succeed should the arbitration proceed to a hearing, we recognise that an early settlement would be in the interests of both parties in reducing uncertainty, in avoiding further costs, and in enabling both parties to deploy their staffs to more productive activities. With these aims in mind I am authorised by my Board to offer the following:

1. This company will, within 14 days of your company's acceptance of this offer, pay the sum of £2,760,000 (two million, seven hundred and sixty thousand pounds) in full and final settlement of all claims and counterclaims arising from the contract between us for the construction of the bridge over the river Avon, including interest.

2. In addition UFC will pay Longspan Bridgebuilders' costs reasonably incurred in the arbitration up to the date of this offer: such costs, if not agreed, to be determined by the Arbitrator.

3. UFC will also pay the Arbitrator's costs.

This offer is made without admission of liability and is open for acceptance within 21 days of its date of issue, namnely 13 March 2001. If it is not accepted it will be privileged from disclosure in evidence during the hearing of the substantive issues, but will be brought to the notice of the Arbitrator before he deals with costs.

Yours faithfully

J Robertson
Managing Director

From:        Universal Finance Corporation PLC

To:          Longspan Bridgebuilders PLC

18 January 1998

Dear Sirs

**Bridge over the River Avon**

1.  Subject to the granting of necessary planning consents it is the intention of Universal Finance Corporation (UFC) to accept your tender dated 20 December 1997 for the above project.

2.  This letter does not constitute an acceptance of your tender, but it authorises you to proceed with the following preliminary works; subject to the right of UFC to rescind this authority at any time:
    2.1  The preparation of designs and drawings for temporary works
    2.2  The erection of site offices, workshops and stores
    2.3  The construction of temporary access roads to and on the site
    2.4  The ordering of temporary and permanent works materials.

3.  Payment will be made for authorised work done and materials delivered to the site on a *quantum meruit* basis subject to your complying with the terms of this letter and of the tender documents; and subject to retention as provided for therein. Such payment will be credited against your entitlement to payment under any contract for the above works that may come into existence.

4.  All goods and materials delivered to the site under the terms of this letter shall vest in UFC and shall be stored safely on site until incorporated in the works or until the rescission of this authority. You shall provide insurances against all risks as specified in the tender documents while such goods and materials remain in your care and shall in the event of rescission of this authority deliver the goods and materials to UFC.

5.  In the event that UFC exercises its power under paragraph 2 above to rescind this authority UFC shall within one month of rescission pay to you all monies payable under the terms of this letter together with all retention monies held by UFC.

6.  Any dispute or difference between the parties that may arise from this letter and your acceptance of its terms shall be determined in accordance with the provisions of clause 66 of the ICE Conditions of Contract, whether or not the contract envisaged therein comes into existence.

7.  Please confirm your acceptance of these terms by signing and returning the enclosed copy of this letter.

Yours faithfully

# APPENDIX B
# THE ARBITRATION ACT 1996

## ARRANGEMENT OF SECTIONS

PART I

ARBITRATION PURSUANT TO AN ARBITRATION AGREEMENT

*Introductory*

# Arbitration Act 1996

An Act to restate and improve the law relating to arbitration pursuant to an arbitration agreement; to make other provision relating to arbitration and arbitration awards; and for connected purposes. [17th June 1996]

## PART I

ARBITRATION PURSUANT TO AN ARBITRATION AGREEMENT

### Introductory

1. The provisions of this Part are founded on the following principles, and shall be construed accordingly —

General principles.

  (a) the object of arbitration is to obtain the fair resolution of disputes by an impartial tribunal without unnecessary delay or expense;

  (b) the parties should be free to agree how their disputes are resolved, subject only to such safeguards as are necessary in the public interest;

  (c) in matters governed by this Part the court should not intervene except as provided by this Part.

2. — (1) The provisions of this Part apply where the seat of the arbitration is in England and Wales or Northern Ireland.

Scope of application of provisions.

(2) The following sections apply even if the seat of the arbitration is outside England and Wales or Northern Ireland or no seat has been designated or determined —

  (a) sections 9 to 11 (stay of legal proceedings, &c.), and

  (b) section 66 (enforcement of arbitral awards).

(3) The powers conferred by the following sections apply even if the seat of the arbitration is outside England and Wales or Northern Ireland or no seat has been designated or determined —

  (a) section 43 (securing the attendance of witnesses), and

  (b) section 44 (court powers exercisable in support of arbitral proceedings);

but the court may refuse to exercise any such power if, in the opinion of the court, the fact that the seat of the arbitration is outside England and Wales or Northern Ireland, or that when designated or determined the seat is likely to be outside England and Wales or Northern Ireland, makes it inappropriate to do so.

(4) The court may exercise a power conferred by any provision of this Part not mentioned in subsection (2) or (3) for the purpose of supporting the arbitral process where —

  (a) no seat of the arbitration has been designated or determined, and

  (b) by reason of a connection with England and Wales or Northern Ireland the court is satisfied that it is appropriate to do so.

(5) Section 7 (separability of arbitration agreement) and section 8 (death of a party) apply where the law applicable to the arbitration agreement is the law of England and Wales or Northern Ireland even if the seat of the

189

arbitration is outside England and Wales or Northern Ireland or has not been designated or determined.

**The seat of the arbitration.**

**3.** In this Part "the seat of the arbitration" means the juridical seat of the arbitration designated—
  (a) by the parties to the arbitration agreement, or
  (b) by any arbitral or other institution or person vested by the parties with powers in that regard, or
  (c) by the arbitral tribunal if so authorised by the parties,
or determined, in the absence of any such designation, having regard to the parties' agreement and all the relevant circumstances.

**Mandatory and non-mandatory provisions.**

**4.**—(1) The mandatory provisions of this Part are listed in Schedule 1 and have effect notwithstanding any agreement to the contrary.

(2) The other provisions of this Part (the "non-mandatory provisions") allow the parties to make their own arrangements by agreement but provide rules which apply in the absence of such agreement.

(3) The parties may make such arrangements by agreeing to the application of institutional rules or providing any other means by which a matter may be decided.

(4) It is immaterial whether or not the law applicable to the parties' agreement is the law of England and Wales or, as the case may be, Northern Ireland.

(5) The choice of a law other than the law of England and Wales or Northern Ireland as the applicable law in respect of a matter provided for by a non-mandatory provision of this Part is equivalent to an agreement making provision about that matter.
  For this purpose an applicable law determined in accordance with the parties' agreement, or which is objectively determined in the absence of any express or implied choice, shall be treated as chosen by the parties.

**Agreements to be in writing.**

**5.**—(1) The provisions of this Part apply only where the arbitration agreement is in writing, and any other agreement between the parties as to any matter is effective for the purposes of this Part only if in writing.
  The expressions "agreement", "agree" and "agreed" shall be construed accordingly.

(2) There is an agreement in writing—
  (a) if the agreement is made in writing (whether or not it is signed by the parties),
  (b) if the agreement is made by exchange of communications in writing, or
  (c) if the agreement is evidenced in writing.

(3) Where parties agree otherwise than in writing by reference to terms which are in writing, they make an agreement in writing.

(4) An agreement is evidenced in writing if an agreement made otherwise than in writing is recorded by one of the parties, or by a third party, with the authority of the parties to the agreement.

(5) An exchange of written submissions in arbitral or legal proceedings in which the existence of an agreement otherwise than in writing is alleged by one party against another party and not denied by the other party in his response constitutes as between those parties an agreement in writing to the effect alleged.

(6) References in this Part to anything being written or in writing include its being recorded by any means.

### The arbitration agreement

**6.**—(1) In this Part an "arbitration agreement" means an agreement to submit to arbitration present or future disputes (whether they are contractual or not).

*Definition of arbitration agreement.*

(2) The reference in an agreement to a written form of arbitration clause or to a document containing an arbitration clause constitutes an arbitration agreement if the reference is such as to make that clause part of the agreement.

**7.** Unless otherwise agreed by the parties, an arbitration agreement which forms or was intended to form part of another agreement (whether or not in writing) shall not be regarded as invalid, non-existent or ineffective because that other agreement is invalid, or did not come into existence or has become ineffective, and it shall for that purpose be treated as a distinct agreement.

*Separability of arbitration agreement.*

**8.**—(1) Unless otherwise agreed by the parties, an arbitration agreement is not discharged by the death of a party and may be enforced by or against the personal representatives of that party.

*Whether agreement discharged by death of a party.*

(2) Subsection (1) does not affect the operation of any enactment or rule of law by virtue of which a substantive right or obligation is extinguished by death.

### Stay of legal proceedings

**9.**—(1) A party to an arbitration agreement against whom legal proceedings are brought (whether by way of claim or counterclaim) in respect of a matter which under the agreement is to be referred to arbitration may (upon notice to the other parties to the proceedings) apply to the court in which the proceedings have been brought to stay the proceedings so far as they concern that matter.

*Stay of legal proceedings.*

(2) An application may be made notwithstanding that the matter is to be referred to arbitration only after the exhaustion of other dispute resolution procedures.

(3) An application may not be made by a person before taking the appropriate procedural step (if any) to acknowledge the legal proceedings against him or after he has taken any step in those proceedings to answer the substantive claim.

(4) On an application under this section the court shall grant a stay unless satisfied that the arbitration agreement is null and void, inoperative, or incapable of being performed.

(5) If the court refuses to stay the legal proceedings, any provision that an award is a condition precedent to the bringing of legal proceedings in respect of any matter is of no effect in relation to those proceedings.

**Reference of interpleader issue to arbitration.**
**10.**—(1) Where in legal proceedings relief by way of interpleader is granted and any issue between the claimants is one in respect of which there is an arbitration agreement between them, the court granting the relief shall direct that the issue be determined in accordance with the agreement unless the circumstances are such that proceedings brought by a claimant in respect of the matter would not be stayed.

(2) Where subsection (1) applies but the court does not direct that the issue be determined in accordance with the arbitration agreement, any provision that an award is a condition precedent to the bringing of legal proceedings in respect of any matter shall not affect the determination of that issue by the court.

**Retention of security where Admiralty proceedings stayed.**
**11.**—(1) Where Admiralty proceedings are stayed on the ground that the dispute in question should be submitted to arbitration, the court granting the stay may, if in those proceedings property has been arrested or bail or other security has been given to prevent or obtain release from arrest—

(a) order that the property arrested be retained as security for the satisfaction of any award given in the arbitration in respect of that dispute, or

(b) order that the stay of those proceedings be conditional on the provision of equivalent security for the satisfaction of any such award.

(2) Subject to any provision made by rules of court and to any necessary modifications, the same law and practice shall apply in relation to property retained in pursuance of an order as would apply if it were held for the purposes of proceedings in the court making the order.

**Power of court to extend time for beginning arbitral proceedings, &c.**

*Commencement of arbitral proceedings*

**12.**—(1) Where an arbitration agreement to refer future disputes to arbitration provides that a claim shall be barred, or the claimant's right extinguished, unless the claimant takes within a time fixed by the agreement some step—

    (a) to begin arbitral proceedings, or

    (b) to begin other dispute resolution procedures which must be exhausted before arbitral proceedings can be begun,

the court may by order extend the time for taking that step.

(2) Any party to the arbitration agreement may apply for such an order (upon notice to the other parties), but only after a claim has arisen and after exhausting any available arbitral process for obtaining an extension of time.

(3) The court shall make an order only if satisfied—

    (a) that the circumstances are such as were outside the reasonable contemplation of the parties when they agreed the provision in question, and that it would be just to extend the time, or

    (b) that the conduct of one party makes it unjust to hold the other party to the strict terms of the provision in question.

(4) The court may extend the time for such period and on such terms as it thinks fit, and may do so whether or not the time previously fixed (by agreement or by a previous order) has expired.

(5) An order under this section does not affect the operation of the Limitation Acts (see section 13).

(6) The leave of the court is required for any appeal from a decision of the court under this section.

**13.**—(1) The Limitation Acts apply to arbitral proceedings as they apply to legal proceedings.     *Application of Limitation Acts.*

(2) The court may order that in computing the time prescribed by the Limitation Acts for the commencement of proceedings (including arbitral proceedings) in respect of a dispute which was the subject matter—

    (a) of an award which the court orders to be set aside or declares to be of no effect, or

    (b) of the affected part of an award which the court orders to be set aside in part, or declares to be in part of no effect,

the period between the commencement of the arbitration and the date of the order referred to in paragraph (a) or (b) shall be excluded.

(3) In determining for the purposes of the Limitation Acts when a cause of action accrued, any provision that an award is a condition precedent to the bringing of legal proceedings in respect of a matter to which an arbitration agreement applies shall be disregarded.

(4) In this Part "the Limitation Acts" means—

    (a) in England and Wales, the Limitation Act 1980, the Foreign Limitation Periods Act 1984 and any other enactment (whenever passed) relating to the limitation of actions;

(b) in Northern Ireland, the Limitation (Northern Ireland) Order 1989, the Foreign Limitation Periods (Northern Ireland) Order 1985 and any other enactment (whenever passed) relating to the limitation of actions.

**Commencement of arbitral proceedings.**

14. — (1) The parties are free to agree when arbitral proceedings are to be regarded as commenced for the purposes of this Part and for the purposes of the Limitation Acts.

(2) If there is no such agreement the following provisions apply.

(3) Where the arbitrator is named or designated in the arbitration agreement, arbitral proceedings are commenced in respect of a matter when one party serves on the other party or parties a notice in writing requiring him or them to submit that matter to the person so named or designated.

(4) Where the arbitrator or arbitrators are to be appointed by the parties, arbitral proceedings are commenced in respect of a matter when one party serves on the other party or parties notice in writing requiring him or them to appoint an arbitrator or to agree to the appointment of an arbitrator in respect of that matter.

(5) Where the arbitrator or arbitrators are to be appointed by a person other than a party to the proceedings, arbitral proceedings are commenced in respect of a matter when one party gives notice in writing to that person requesting him to make the appointment in respect of that matter.

*The arbitral tribunal*

**The arbitral tribunal.**

15. — (1) The parties are free to agree on the number of arbitrators to form the tribunal and whether there is to be a chairman or umpire.

(2) Unless otherwise agreed by the parties, an agreement that the number of arbitrators shall be two or any other even number shall be understood as requiring the appointment of an additional arbitrator as chairman of the tribunal.

(3) If there is no agreement as to the number of arbitrators, the tribunal shall consist of a sole arbitrator.

**Procedure for appointment of arbitrators.**

16. — (1) The parties are free to agree on the procedure for appointing the arbitrator or arbitrators, including the procedure for appointing any chairman or umpire.

(2) If or to the extent that there is no such agreement, the following provisions apply.

(3) If the tribunal is to consist of a sole arbitrator, the parties shall jointly appoint the arbitrator not later than 28 days after service of a request in writing by either party to do so.

(4) If the tribunal is to consist of two arbitrators, each party shall appoint one arbitrator not later than 14 days after service of a request in writing by either party to do so.

(5) If the tribunal is to consist of three arbitrators —
  (a) each party shall appoint one arbitrator not later than 14 days after service of a request in writing by either party to do so, and
  (b) the two so appointed shall forthwith appoint a third arbitrator as the chairman of the tribunal.

(6) If the tribunal is to consist of two arbitrators and an umpire —
  (a) each party shall appoint one arbitrator not later than 14 days after service of a request in writing by either party to do so, and
  (b) the two so appointed may appoint an umpire at any time after they themselves are appointed and shall do so before any substantive hearing or forthwith if they cannot agree on a matter relating to the arbitration.

(7) In any other case (in particular, if there are more than two parties) section 18 applies as in the case of a failure of the agreed appointment procedure.

17. — (1) Unless the parties otherwise agree, where each of two parties to an arbitration agreement is to appoint an arbitrator and one party ("the party in default") refuses to do so, or fails to do so within the time specified, the other party, having duly appointed his arbitrator, may give notice in writing to the party in default that he proposes to appoint his arbitrator to act as sole arbitrator. *Power in case of default to appoint sole arbitrator.*

(2) If the party in default does not within 7 clear days of that notice being given —
  (a) make the required appointment, and
  (b) notify the other party that he has done so,
the other party may appoint his arbitrator as sole arbitrator whose award shall be binding on both parties as if he had been so appointed by agreement.

(3) Where a sole arbitrator has been appointed under subsection (2), the party in default may (upon notice to the appointing party) apply to the court which may set aside the appointment.

(4) The leave of the court is required for any appeal from a decision of the court under this section.

18. — (1) The parties are free to agree what is to happen in the event of a failure of the procedure for the appointment of the arbitral tribunal. *Failure of appointment procedure.*
  There is no failure if an appointment is duly made under section 17 (power in case of default to appoint sole arbitrator), unless that appointment is set aside.

(2) If or to the extent that there is no such agreement any party to the arbitration agreement may (upon notice to the other parties) apply to the court to exercise its powers under this section.

(3) Those powers are—
   (a) to give directions as to the making of any necessary appointments;
   (b) to direct that the tribunal shall be constituted by such appoint-ments (or any one or more of them) as have been made;
   (c) to revoke any appointments already made;
   (d) to make any necessary appointments itself.

(4) An appointment made by the court under this section has effect as if made with the agreement of the parties.

(5) The leave of the court is required for any appeal from a decision of the court under this section.

**Court to have regard to agreed qualifications.**
   **19.** In deciding whether to exercise, and in considering how to exercise, any of its powers under section 16 (procedure for appointment of arbi-trators) or section 18 (failure of appointment procedure), the court shall have due regard to any agreement of the parties as to the qualifications required of the arbitrators.

**Chairman.**
   **20.**—(1) Where the parties have agreed that there is to be a chairman, they are free to agree what the functions of the chairman are to be in relation to the making of decisions, orders and awards.

(2) If or to the extent that there is no such agreement, the following provisions apply.

(3) Decisions, orders and awards shall be made by all or a majority of the arbitrators (including the chairman).

(4) The view of the chairman shall prevail in relation to a decision, order or award in respect of which there is neither unanimity nor a majority under subsection (3).

**Umpire.**
   **21.**—(1) Where the parties have agreed that there is to be an umpire, they are free to agree what the functions of the umpire are to be, and in particular—
   (a) whether he is to attend the proceedings, and
   (b) when he is to replace the other arbitrators as the tribunal with power to make decisions, orders and awards.

(2) If or to the extent that there is no such agreement, the following provisions apply.

(3) The umpire shall attend the proceedings and be supplied with the same documents and other materials as are supplied to the other arbi-trators.

(4) Decisions, orders and awards shall by made by the other arbitrators unless and until they cannot agree on a matter relating to the arbitration.

In that event they shall forthwith give notice in writing to the parties and the umpire, whereupon the umpire shall replace them as the tribunal with power to make decisions, orders and awards as if he were sole arbitrator.

(5) If the arbitrators cannot agree but fail to give notice of that fact, or if any of them fails to join in the giving of notice, any party to the arbitral proceedings may (upon notice to the other parties and to the tribunal) apply to the court which may order that the umpire shall replace the other arbitrators as the tribunal with power to make decisions, orders and awards as if he were sole arbitrator.

(6) The leave of the court is required for any appeal from a decision of the court under this section.

**22.** — (1) Where the parties agree that there shall be two or more arbitrators with no chairman or umpire, the parties are free to agree how the tribunal is to make decisions, orders and awards. *Decision-making where no chairman or umpire.*

(2) If there is no such agreement, decisions, orders and awards shall be made by all or a majority of the arbitrators.

**23.** — (1) The parties are free to agree in what circumstances the authority of an arbitrator may be revoked. *Revocation of arbitrator's authority.*

(2) If or to the extent that there is no such agreement the following provisions apply.

(3) The authority of an arbitrator may not be revoked except —
  (a) by the parties acting jointly, or
  (b) by an arbitral or other institution or person vested by the parties with powers in that regard.

(4) Revocation of the authority of an arbitrator by the parties acting jointly must be agreed in writing unless the parties also agree (whether or not in writing) to terminate the arbitration agreement.

(5) Nothing in this section affects the power of the court —
  (a) to revoke an appointment under section 18 (powers exercisable in case of failure of appointment procedure), or
  (b) to remove an arbitrator on the grounds specified in section 24.

**24.** — (1) A party to arbitral proceedings may (upon notice to the other parties, to the arbitrator concerned and to any other arbitrator) apply to the court to remove an arbitrator on any of the following grounds — *Power of court to remove arbitrator.*
  (a) that circumstances exist that give rise to justifiable doubts as to his impartiality;
  (b) that he does not possess the qualifications required by the arbitration agreement;

(c) that he is physically or mentally incapable of conducting the proceedings or there are justifiable doubts as to his capacity to do so;

(d) that he has refused or failed —
    (i) properly to conduct the proceedings, or
    (ii) to use all reasonable despatch in conducting the proceedings or making an award,

and that substantial injustice has been or will be caused to the applicant.

(2) If there is an arbitral or other institution or person vested by the parties with power to remove an arbitrator, the court shall not exercise its power of removal unless satisfied that the applicant has first exhausted any available recourse to that institution or person.

(3) The arbitral tribunal may continue the arbitral proceedings and make an award while an application to the court under this section is pending.

(4) Where the court removes an arbitrator, it may make such order as it thinks fit with respect to his entitlement (if any) to fees or expenses, or the repayment of any fees or expenses already paid.

(5) The arbitrator concerned is entitled to appear and be heard by the court before it makes any order under this section.

(6) The leave of the court is required for any appeal from a decision of the court under this section.

**Resignation of arbitrator.**

**25.** — (1) The parties are free to agree with an arbitrator as to the consequences of his resignation as regards —
    (a) his entitlement (if any) to fees or expenses, and
    (b) any liability thereby incurred by him.

(2) If or to the extent that there is no such agreement the following provisions apply.

(3) An arbitrator who resigns his appointment may (upon notice to the parties) apply to the court —
    (a) to grant him relief from any liability thereby incurred by him, and
    (b) to make such order as it thinks fit with respect to his entitlement (if any) to fees or expenses or the repayment of any fees or expenses already paid.

(4) If the court is satisfied that in all the circumstances it was reasonable for the arbitrator to resign, it may grant such relief as is mentioned in subsection (3)(a) on such terms as it thinks fit.

(5) The leave of the court is required for any appeal from a decision of the court under this section.

26.—(1) The authority of an arbitrator is personal and ceases on his death.

(2) Unless otherwise agreed by the parties, the death of the person by whom an arbitrator was appointed does not revoke the arbitrator's authority.

27.—(1) Where an arbitrator ceases to hold office, the parties are free to agree—

(a) whether and if so how the vacancy is to be filled,
(b) whether and if so to what extent the previous proceedings should stand, and
(c) what effect (if any) his ceasing to hold office has on any appointment made by him (alone or jointly).

(2) If or to the extent that there is no such agreement, the following provisions apply.

(3) The provisions of sections 16 (procedure for appointment of arbitrators) and 18 (failure of appointment procedure) apply in relation to the filling of the vacancy as in relation to an original appointment.

(4) The tribunal (when reconstituted) shall determine whether and if so to what extent the previous proceedings should stand.
This does not affect any right of a party to challenge those proceedings on any ground which had arisen before the arbitrator ceased to hold office.

(5) His ceasing to hold office does not affect any appointment by him (alone or jointly) of another arbitrator, in particular any appointment of a chairman or umpire.

28.—(1) The parties are jointly and severally liable to pay to the arbitrators such reasonable fees and expenses (if any) as are appropriate in the circumstances.

(2) Any party may apply to the court (upon notice to the other parties and to the arbitrators) which may order that the amount of the arbitrators' fees and expenses shall be considered and adjusted by such means and upon such terms as it may direct.

(3) If the application is made after any amount has been paid to the arbitrators by way of fees or expenses, the court may order the repayment of such amount (if any) as is shown to be excessive, but shall not do so unless it is shown that it is reasonable in the circumstances to order repayment.

(4) The above provisions have effect subject to any order of the court under section 24(4) or 25(3)(b) (order as to entitlement to fees or expenses in case of removal or resignation of arbitrator).

(5) Nothing in this section affects any liability of a party to any other

party to pay all or any of the costs of the arbitration (see sections 59 to 65) or any contractual right of an arbitrator to payment of his fees and expenses.

(6) In this section references to arbitrators include an arbitrator who has ceased to act and an umpire who has not replaced the other arbitrators.

Immunity of arbitrator.

**29.** — (1) An arbitrator is not liable for anything done or omitted in the discharge or purported discharge of his functions as arbitrator unless the act or omission is shown to have been in bad faith.

(2) Subsection (1) applies to an employee or agent of an arbitrator as it applies to the arbitrator himself.

(3) This section does not affect any liability incurred by an arbitrator by reason of his resigning (but see section 25).

*Jurisdiction of the arbitral tribunal*

Competence of tribunal to rule on its own jurisdiction.

**30.** — (1) Unless otherwise agreed by the parties, the arbitral tribunal may rule on its own substantive jurisdiction, that is, as to —
  (a) whether there is a valid arbitration agreement,
  (b) whether the tribunal is properly constituted, and
  (c) what matters have been submitted to arbitration in accordance with the arbitration agreement.

(2) Any such ruling may be challenged by any available arbitral process of appeal or review or in accordance with the provisions of this Part.

Objection to substantive jurisdiction of tribunal.

**31.** — (1) An objection that the arbitral tribunal lacks substantive jurisdiction at the outset of the proceedings must be raised by a party not later than the time he takes the first step in the proceedings to contest the merits of any matter in relation to which he challenges the tribunal's jurisdiction.
  A party is not precluded from raising such an objection by the fact that he has appointed or participated in the appointment of an arbitrator.

(2) Any objection during the course of the arbitral proceedings that the arbitral tribunal is exceeding its substantive jurisdiction must be made as soon as possible after the matter alleged to be beyond its jurisdiction is raised.

(3) The arbitral tribunal may admit an objection later than the time specified in subsection (1) or (2) if it considers the delay justified.

(4) Where an objection is duly taken to the tribunal's substantive jurisdiction and the tribunal has power to rule on its own jurisdiction, it may —
  (a) rule on the matter in an award as to jurisdiction, or
  (b) deal with the objection in its award on the merits.
If the parties agree which of these courses the tribunal should take, the tribunal shall proceed accordingly.

(5) The tribunal may in any case, and shall if the parties so agree, stay proceedings whilst an application is made to the court under section 32 (determination of preliminary point of jurisdiction).

**32.**—(1) The court may, on the application of a party to arbitral proceedings (upon notice to the other parties), determine any question as to the substantive jurisdiction of the tribunal.

A party may lose the right to object (see section 73).

> *Determination of preliminary point of jurisdiction.*

(2) An application under this section shall not be considered unless—
- (a) it is made with the agreement in writing of all the other parties to the proceedings, or
- (b) it is made with the permission of the tribunal and the court is satisfied—
  - (i) that the determination of the question is likely to produce substantial savings in costs,
  - (ii) that the application was made without delay, and
  - (iii) that there is good reason why the matter should be decided by the court.

(3) An application under this section, unless made with the agreement of all the other parties to the proceedings, shall state the grounds on which it is said that the matter should be decided by the court.

(4) Unless otherwise agreed by the parties, the arbitral tribunal may continue the arbitral proceedings and make an award while an application to the court under this section is pending.

(5) Unless the court gives leave, no appeal lies from a decision of the court whether the conditions specified in subsection (2) are met.

(6) The decision of the court on the question of jurisdiction shall be treated as a judgment of the court for the purposes of an appeal.

But no appeal lies without the leave of the court which shall not be given unless the court considers that the question involves a point of law which is one of general importance or is one which for some other special reason should be considered by the Court of Appeal.

## The arbitral proceedings

**33.**—(1) The tribunal shall—
- (a) act fairly and impartially as between the parties, giving each party a reasonable opportunity of putting his case and dealing with that of his opponent, and
- (b) adopt procedures suitable to the circumstances of the particular case, avoiding unnecessary delay or expense, so as to provide a fair means for the resolution of the matters falling to be determined.

> *General duty of the tribunal.*

(2) The tribunal shall comply with that general duty in conducting the arbitral proceedings, in its decisions on matters of procedure and evidence and in the exercise of all other powers conferred on it.

**Procedural and evidential matters.**

**34.** — (1) It shall be for the tribunal to decide all procedural and evidential matters, subject to the right of the parties to agree any matter.

(2) Procedural and evidential matters include —
- (a) when and where any part of the proceedings is to be held;
- (b) the language or languages to be used in the proceedings and whether translations of any relevant documents are to be supplied;
- (c) whether any and if so what form of written statements of claim and defence are to be used, when these should be supplied and the extent to which such statements can be later amended;
- (d) whether any and if so which documents or classes of documents should be disclosed between and produced by the parties and at what stage;
- (e) whether any and if so what questions should be put to and answered by the respective parties and when and in what form this should be done;
- (f) whether to apply strict rules of evidence (or any other rules) as to the admissibility, relevance or weight of any material (oral, written or other) sought to be tendered on any matters of fact or opinion, and the time, manner and form in which such material should be exchanged and presented;
- (g) whether and to what extent the tribunal should itself take the initiative in ascertaining the facts and the law;
- (h) whether and to what extent there should be oral or written evidence or submissions.

(3) The tribunal may fix the time within which any directions given by it are to be complied with, and may if it thinks fit extend the time so fixed (whether or not it has expired).

**Consolidation of proceedings and concurrent hearings.**

**35.** — (1) The parties are free to agree —
- (a) that the arbitral proceedings shall be consolidated with other arbitral proceedings, or
- (b) that concurrent hearings shall be held, on such terms as may be agreed.

(2) Unless the parties agree to confer such power on the tribunal, the tribunal has no power to order consolidation of proceedings or concurrent hearings.

**Legal or other representation.**

**36.** Unless otherwise agreed by the parties, a party to arbitral proceedings may be represented in the proceedings by a lawyer or other person chosen by him.

Power to appoint
experts, legal
advisers or
assessors.

37.—(1) Unless otherwise agreed by the parties—
   (a) the tribunal may—
      (i) appoint experts or legal advisers to report to it and the parties, or
      (ii) appoint assessors to assist it on technical matters,
   and may allow any such expert, legal adviser or assessor to attend the proceedings; and
   (b) the parties shall be given a reasonable opportunity to comment on any information, opinion or advice offered by any such person.

(2) The fees and expenses of an expert, legal adviser or assessor appointed by the tribunal for which the arbitrators are liable are expenses or the arbitrators for the purposes of this Part.

General powers
exercisable by
the tribunal.

38.—(1) The parties are free to agree on the powers exercisable by the arbitral tribunal for the purposes of and in relation to the proceedings.

(2) Unless otherwise agreed by the parties the tribunal has the following powers.

(3) The tribunal may order a claimant to provide security for the costs of the arbitration.
This power shall not be exercised on the ground that the claimant is—
   (a) an individual ordinarily resident outside the United Kingdom, or
   (b) a corporation or association incorporated or formed under the law of a country outside the United Kingdom, or whose central management and control is exercised outside the United Kingdom.

(4) The tribunal may give directions in relation to any property which is the subject of the proceedings or as to which any question arises in the proceedings, and which is owned by or is in the possession of a party to the proceedings—
   (a) for the inspection, photographing, preservation, custody or detention of the property by the tribunal, an expert or a party, or
   (b) ordering that samples be taken from, or any observation be made of or experiment conducted upon, the property.

(5) The tribunal may direct that a party or witness shall be examined on oath or affirmation, and may for that purpose administer any necessary oath or take any necessary affirmation.

(6) The tribunal may give directions to a party for the preservation for the purposes of the proceedings of any evidence in his custody or control.

Power to make
provisional
awards.

39.—(1) The parties are free to agree that the tribunal shall have power to order on a provisional basis any relief which it would have power to grant in a final award.

(2) This includes, for instance, making—

(a) a provisional order for the payment of money or the disposition of property as between the parties, or

(b) an order to make an interim payment on account of the costs of the arbitration.

(3) Any such order shall be subject to the tribunal's final adjudication; and the tribunal's final award, on the merits or as to costs, shall take account of any such order.

(4) Unless the parties agree to confer such power on the tribunal, the tribunal has no such power.

This does not affect its powers under section 47 (awards on different issues, &c.).

General duty of parties.

**40.** – (1) The parties shall do all things necessary for the proper and expeditious conduct of the arbitral proceedings.

(2) This includes –

(a) complying without delay with any determination of the tribunal as to procedural or evidential matters, or with any order or directions of the tribunal, and

(b) where appropriate, taking without delay any necessary steps to obtain a decision of the court on a preliminary question of jurisdiction or law (see sections 32 and 45).

Powers of tribunal in case of party's default.

**41.** – (1) The parties are free to agree on the powers of the tribunal in case of a party's failure to do something necessary for the proper and expeditious conduct of the arbitration.

(2) Unless otherwise agreed by the parties, the following provisions apply.

(3) If the tribunal is satisfied that there has been inordinate and inexcusable delay on the part of the claimant in pursuing his claim and that the delay –

(a) gives rise, or is likely to give rise, to a substantial risk that it is not possible to have a fair resolution of the issues in that claim, or

(b) has caused, or is likely to cause, serious prejudice to the respondent,

the tribunal may make an award dismissing the claim.

(4) If without showing sufficient cause a party –

(a) fails to attend or be represented at an oral hearing of which due notice was given, or

(b) where matters are to be dealt with in writing, fails after due notice to submit written evidence or make written submissions,

the tribunal may continue the proceedings in the absence of that party or, as the case may be, without any written evidence or submissions on his behalf, and may make an award on the basis of the evidence before it.

(5) If without showing sufficient cause a party fails to comply with any order or directions of the tribunal, the tribunal may make a peremptory order to the same effect, prescribing such time for compliance with it as the tribunal considers appropriate.

(6) If a claimant fails to comply with a peremptory order of the tribunal to provide security for costs, the tribunal may make an award dismissing his claim.

(7) If a party fails to comply with any other kind of peremptory order, then, without prejudice to section 42 (enforcement by court of tribunal's peremptory orders), the tribunal may do any of the following—
  (a) direct that the party in default shall not be entitled to rely upon any allegation or material which was the subject matter of the order;
  (b) draw such adverse inferences from the act of non-compliance as the circumstances justify;
  (c) proceed to an award on the basis of such materials as have been properly provided to it;
  (d) make such order as it thinks fit as to the payment of costs of the arbitration incurred in consequence of the non-compliance.

### Powers of court in relation to arbitral proceedings

**42.**—(1) Unless otherwise agreed by the parties, the court may make an order requiring a party to comply with a peremptory order made by the tribunal.

**Enforcement of peremptory orders of tribunal.**

(2) An application for an order under this section may be made—
  (a) by the tribunal (upon notice to the parties),
  (b) by a party to the arbitral proceedings with the permission of the tribunal (and upon notice to the other parties), or
  (c) where the parties have agreed that the powers of the court under this section shall be available.

(3) The court shall not act unless it is satisfied that the applicant has exhausted any available arbitral process in respect of failure to comply with the tribunal's order.

(4) No order shall be made under this section unless the court is satisfied that the person to whom the tribunal's order was directed has failed to comply with it within the time prescribed in the order or, if no time was prescribed, within a reasonable time.

(5) The leave of the court is required for any appeal from a decision of the court under this section.

**43.**—(1) A party to arbitral proceedings may use the same court procedures as are available in relation to legal proceedings to secure the attendance before the tribunal of a witness in order to give oral testimony or to produce documents or other material evidence.

**Securing the attendance of witnesses.**

(2) This may only be done with the permission of the tribunal or the agreement of the other parties.

(3) The court procedures may only be used if—
   (a) the witness is in the United Kingdom, and
   (b) the arbitral proceedings are being conducted in England and Wales or, as the case may be, Northern Ireland.

(4) A person shall not be compelled by virtue of this section to produce any document or other material evidence which he could not be compelled to produce in legal proceedings.

**Court powers exercisable in support of arbitral proceedings.**

**44.**—(1) Unless otherwise agreed by the parties, the court has for the purposes of and in relation to arbitral proceedings the same power of making orders about the matters listed below as it has for the purposes of and in relation to legal proceedings.

(2) Those matters are—
   (a) the taking of the evidence of witnesses;
   (b) the preservation of evidence;
   (c) making orders relating to property which is the subject of the proceedings or as to which any question arises in the proceedings—
      (i) for the inspection, photographing, preservation, custody or detention of the property, or
      (ii) ordering that samples be taken from, or any observation be made of or experiment conducted upon, the property;
      and for that purpose authorising any person to enter any premises in the possession or control of a party to the arbitration;
   (d) the sale of any goods the subject of the proceedings;
   (e) the granting of an interim injunction or the appointment of a receiver.

(3) If the case is one of urgency, the court may, on the application of a party or proposed party to the arbitral proceedings, make such orders as it thinks necessary for the purpose of preserving evidence or assets.

(4) If the case is not one of urgency, the court shall act only on the application of a party to the arbitral proceedings (upon notice to the other parties and to the tribunal) made with the permission of the tribunal or the agreement in writing of the other parties.

(5) In any case the court shall act only if or to the extent that the arbitral tribunal, and any arbitral or other institution or person vested by the parties with power in that regard, has no power or is unable for the time being to act effectively.

(6) If the court so orders, an order made by it under this section shall cease to have effect in whole or in part on the order of the tribunal or of any such arbitral or other institution or person having power to act in relation to the subject-matter of the order.

(7) The leave of the court is required for any appeal from a decision of the court under this section.

**45.**—(1) Unless otherwise agreed by the parties, the court may on the application of a party to arbitral proceedings (upon notice to the other parties) determine any question of law arising in the course of the proceedings which the court is satisfied substantially affects the rights of one or more of the parties.

Determination of preliminary point of law.

An agreement to dispense with reasons for the tribunal's award shall be considered an agreement to exclude the court's jurisdiction under this section.

(2) An application under this section shall not be considered unless—
    (a) it is made with the agreement of all the other parties to the proceedings, or
    (b) it is made with the permission of the tribunal and the court is satisfied—
        (i) that the determination of the question is likely to produce substantial savings in costs, and
        (ii) that the application was made without delay.

(3) The application shall identify the question of law to be determined and, unless made with the agreement of all the other parties to the proceedings, shall state the grounds on which it is said that the question should be decided by the court.

(4) Unless otherwise agreed by the parties, the arbitral tribunal may continue the arbitral proceedings and make an award while an application to the court under this section is pending.

(5) Unless the court gives leave, no appeal lies from a decision of the court whether the conditions specified in subsection (2) are met.

(6) The decision of the court on the question of law shall be treated as a judgment of the court for the purposes of an appeal.

But no appeal lies without the leave of the court which shall not be given unless the court considers that the question is one of general importance, or is one which for some other special reason should be considered by the Court of Appeal.

### The award

**46.**—(1) The arbitral tribunal shall decide the dispute—
    (a) in accordance with the law chosen by the parties as applicable to the substance of the dispute, or
    (b) if the parties so agree, in accordance with such other considerations as are agreed by them or determined by the tribunal.

Rules applicable to substance of dispute.

(2) For this purpose the choice of the laws of a country shall be understood to refer to the substantive laws of that country and not its conflict of laws rules.

(3) If or to the extent that there is no such choice or agreement, the tribunal shall apply the law determined by the conflict of laws rules which it considers applicable.

**Awards on differrent issues, &c.**
47.—(1) Unless otherwise agreed by the parties, the tribunal may make more than one award at different times on different aspects of the matters to be determined.

(2) The tribunal may, in particular, make an award relating—
(a) to an issue affecting the whole claim, or
(b) to a part only of the claims or cross-claims submitted to it for decision.

(3) If the tribunal does so, it shall specify in its award the issue, or the claim or part of a claim, which is the subject matter of the award.

**Remedies.**
48.—(1) The parties are free to agree on the powers exercisable by the arbitral tribunal as regards remedies.

(2) Unless otherwise agreed by the parties, the tribunal has the following powers.

(3) The tribunal may make a declaration as to any matter to be determined in the proceedings.

(4) The tribunal may order the payment of a sum of money, in any currency.

(5) The tribunal has the same powers as the court—
(a) to order a party to do or refrain from doing anything;
(b) to order specific performance of a contract (other than a contract relating to land);
(c) to order the rectification, setting aside or cancellation of a deed or other document.

**Interest.**
49.—(1) The parties are free to agree on the powers of the tribunal as regards the award of interest.

(2) Unless otherwise agreed by the parties the following provisions apply.

(3) The tribunal may award simple or compound interest from such dates, at such rates and with such rests as it considers meets the justice of the case—
(a) on the whole or part of any amount awarded by the tribunal, in respect of any period up to the date of the award;
(b) on the whole or part of any amount claimed in the arbitration and outstanding at the commencement of the arbitral proceedings but paid before the award was made, in respect of any period up to the date of payment.

(4) The tribunal may award simple or compound interest from the date of the award (or any later date) until payment, at such rates and with such rests as it considers meets the justice of the case, on the outstanding amount of any award (including any award of interest under subsection (3) and any award as to costs).

(5) References in this section to an amount awarded by the tribunal include an amount payable in consequence of a declaratory award by the tribunal.

(6) The above provisions do not affect any other power of the tribunal to award interest.

**50.**—(1) Where the time for making an award is limited by or in pursuance of the arbitration agreement, then, unless otherwise agreed by the parties, the court may in accordance with the following provisions by order extend that time.

Extension of time for making award.

(2) An application for an order under this section may be made—
    (a) by the tribunal (upon notice to the parties), or
    (b) by any party to the proceedings (upon notice to the tribunal and the other parties),
but only after exhausting any available arbitral process for obtaining an extension of time.

(3) The court shall only make an order if satisfied that a substantial injustice would otherwise be done.

(4) The court may extend the time for such period and on such terms as it thinks fit, and may do so whether or not the time previously fixed (by or under the agreement or by a previous order) has expired.

(5) The leave of the court is required for any appeal from a decision of the court under this section.

**51.**—(1) If during arbitral proceedings the parties settle the dispute, the following provisions apply unless otherwise agreed by the parties.

Settlement.

(2) The tribunal shall terminate the substantive proceedings and, if so requested by the parties and not objected to by the tribunal, shall record the settlement in the form of an agreed award.

(3) An agreed award shall state that it is an award of the tribunal and shall have the same status and effect as any other award on the merits of the case.

(4) The following provisions of this Part relating to awards (sections 52 to 58) apply to an agreed award.

(5) Unless the parties have also settled the matter of the payment of the costs of the arbitration, the provisions of this Part relating to costs (sections 59 to 65) continue to apply.

Form of award.

**52.**—(1) The parties are free to agree on the form of an award.

(2) If or to the extent that there is no such agreement, the following provisions apply.

(3) The award shall be in writing signed by all the arbitrators or all those assenting to the award.

(4) The award shall contain the reasons for the award unless it is an agreed award or the parties have agreed to dispense with reasons.

(5) The award shall state the seat of the arbitration and the date when the award is made.

Place where award treated as made.

**53.** Unless otherwise agreed by the parties, where the seat of the arbitration is in England and Wales or Northern Ireland, any award in the proceedings shall be treated as made there, regardless of where it was signed, despatched or delivered to any of the parties.

Date of award.

**54.**—(1) Unless otherwise agreed by the parties, the tribunal may decide what is to be taken to be the date on which the award was made.

(2) In the absence of any such decision, the date of the award shall be taken to be the date on which it is signed by the arbitrator or, where more than one arbitrator signs the award, by the last of them.

Notification of award.

**55.**—(1) The parties are free to agree on the requirements as to notification of the award to the parties.

(2) If there is no such agreement, the award shall be notified to the parties by service on them of copies of the award, which shall be done without delay after the award is made.

(3) Nothing in this section affects section 56 (power to withhold award in case of non-payment).

Power to withhold award in case of non-payment.

**56.**—(1) The tribunal may refuse to deliver an award to the parties except upon full payment of the fees and expenses of the arbitrators.

(2) If the tribunal refuses on that ground to deliver an award, a party to the arbitral proceedings may (upon notice to the other parties and the tribunal) apply to the court, which may order that—

    (a) the tribunal shall deliver the award on the payment into court by the applicant of the fees and expenses demanded, or such lesser amount as the court may specify,

    (b) the amount of the fees and expenses properly payable shall be determined by such means and upon such terms as the court may direct, and

    (c) out of the money paid into court there shall be paid out such fees and expenses as may be found to be properly payable and the balance of the money (if any) shall be paid out to the applicant.

(3) For this purpose the amount of fees and expenses properly payable is the amount the applicant is liable to pay under section 28 or any agreement relating to the payment of the arbitrators.

(4) No application to the court may be made where there is any available arbitral process for appeal or review of the amount of the fees or expenses demanded.

(5) References in this section to arbitrators include an arbitrator who has ceased to act and an umpire who has not replaced the other arbitrators.

(6) The above provisions of this section also apply in relation to any arbitral or other institution or person vested by the parties with powers in relation to the delivery of the tribunal's award.

As they so apply, the references to the fees and expenses of the arbitrators shall be construed as including the fees and expenses of that institution or person.

(7) The leave of the court is required for any appeal from a decision of the court under this section.

(8) Nothing in this section shall be construed as excluding an application under section 28 where payment has been made to the arbitrators in order to obtain the award.

**57.** — (1) The parties are free to agree on the powers of the tribunal to correct an award or make an additional award.

Correction of award or additional award.

(2) If or to the extent there is no such agreement, the following provisions apply.

(3) The tribunal may on its own initiative or on the application of a party —
   (a) correct an award so as to remove any clerical mistake or error arising from an accidental slip or omission or clarify or remove any ambiguity in the award, or
   (b) make an additional award in respect of any claim (including a claim for interest or costs) which was presented to the tribunal but was not dealt with in the award.
These powers shall not be exercised without first affording the other parties a reasonable opportunity to make representations to the tribunal.

(4) Any application for the exercise of those powers must be made within 28 days of the date of the award or such longer period as the parties may agree.

(5) Any correction of an award shall be made within 28 days of the date the application was received by the tribunal or, where the correction is made by the tribunal on its own initiative, within 28 days of the date of the award or, in either case, such longer period as the parties may agree.

(6) Any additional award shall be made within 56 days of the date of the original award or such longer period as the parties may agree.

(7) Any correction of an award shall form part of the award.

**Effect of award.**

**58.**—(1) Unless otherwise agreed by the parties, an award made by the tribunal pursuant to an arbitration agreement is final and binding both on the parties and on any persons claiming through or under them.

(2) This does not affect the right of a person to challenge the award by any available arbitral process of appeal or review or in accordance with the provisions of this Part.

*Costs of the arbitration*

**Costs of the arbitration.**

**59.**—(1) References in this Part to the costs of the arbitration are to—
    (a) the arbitrators' fees and expenses,
    (b) the fees and expenses of any arbitral institution concerned, and
    (c) the legal or other costs of the parties.

(2) Any such reference includes the costs of or incidental to any proceedings to determine the amount of the recoverable costs of the arbitration (see section 63).

**Agreement to pay costs in any event.**

**60.** An agreement which has the effect that a party is to pay the whole or part of the costs of the arbitration in any event is only valid if made after the dispute in question has arisen.

**Award of costs.**

**61.**—(1) The tribunal may make an award allocating the costs of the arbitration as between the parties, subject to any agreement of the parties.

(2) Unless the parties otherwise agree, the tribunal shall award costs on the general principle that costs should follow the event except where it appears to the tribunal that in the circumstances this is not appropriate in relation to the whole or part of the costs.

**Effect of agreement or award about costs.**

**62.** Unless the parties otherwise agree, any obligation under an agreement between them as to how the costs of the arbitration are to be borne, or under an award allocating the costs of the arbitration, extends only to such costs as are recoverable.

**The recoverable costs of the arbitration.**

**63.**—(1) The parties are free to agree what costs of the arbitration are recoverable.

(2) If or to the extent there is no such agreement, the following provisions apply.

(3) The tribunal may determine by award the recoverable costs of the arbitration on such basis as it thinks fit.
If it does so, it shall specify—

(a) the basis on which it has acted, and

(b) the items of recoverable costs and the amount referable to each.

(4) If the tribunal does not determine the recoverable costs of the arbitration, any party to the arbitral proceedings may apply to the court (upon notice to the other parties) which may—

(a) determine the recoverable costs of the arbitration on such basis as it thinks fit, or

(b) order that they shall be determined by such means and upon such terms as it may specify.

(5) Unless the tribunal or the court determines otherwise—

(a) the recoverable costs of the arbitration shall be determined on the basis that there shall be allowed a reasonable amount in respect of all costs reasonably incurred, and

(b) any doubt as to whether costs were reasonably incurred or were reasonable in amount shall be resolved in favour of the paying party.

(6) The above provisions have effect subject to section 64 (recoverable fees and expenses of arbitrators).

(7) Nothing in this section affects any right of the arbitrators, any expert, legal adviser or assessor appointed by the tribunal, or any arbitral institution, to payment of their fees and expenses.

**64.** — (1) Unless otherwise agreed by the parties, the recoverable costs of the arbitration shall include in respect of the fees and expenses of the arbitrators only such reasonable fees and expenses as are appropriate in the circumstances.

*Recoverable fees and expenses of arbitrators.*

(2) If there is any question as to what reasonable fees and expenses are appropriate in the circumstances, and the matter is not already before the court on an application under section 63(4), the court may on the application of any party (upon notice to the other parties)—

(a) determine the matter, or

(b) order that it be determined by such means and upon such terms as the court may specify.

(3) Subsection (1) has effect subject to any order of the court under section 24(4) or 25(3)(b) (order as to entitlement to fees or expenses in case of removal or resignation of arbitrator).

(4) Nothing in this section affects any right of the arbitrator to payment of his fees and expenses.

**65.** — (1) Unless otherwise agreed by the parties, the tribunal may direct that the recoverable costs of the arbitration, or of any part of the arbitral proceedings, shall be limited to a specified amount.

*Power to limit recoverable costs.*

(2) Any direction may be made or varied at any stage, but this must be done sufficiently in advance of the incurring of costs to which it relates, or the taking of any steps in the proceedings which may be affected by it, for the limit to be taken into account.

*Powers of the court in relation to award*

**Enforcement of the award.**

**66.** — (1) An award made by the tribunal pursuant to an arbitration agreement may, by leave of the court, be enforced in the same manner as a judgment or order of the court to the same effect.

(2) Where leave is so given, judgment may be entered in terms of the award.

(3) Leave to enforce an award shall not be given where, or to the extent that, the person against whom it is sought to be enforced shows that the tribunal lacked substantive jurisdiction to make the award.
The right to raise such an objection may have been lost (see section 73).

(4) Nothing in this section affects the recognition or enforcement of an award under any other enactment or rule of law, in particular under Part II of the Arbitration Act 1950 (enforcement of awards under Geneva Convention) or the provisions of Part III of this Act relating to the recognition and enforcement of awards under the New York convention or by an action on the award.

**Challenging the award: substantive jurisdiction.**

**67.** — (1) A party to arbitral proceedings may (upon notice to the other parties and to the tribunal) apply to the court —
  (a) challenging any award of the arbitral tribunal as to its substantive jurisdiction; or
  (b) for an order declaring an award made by the tribunal on the merits to be of no effect, in whole or in part, because the tribunal did not have substantive jurisdiction.
A party may lose the right to object (see section 73) and the right to apply is subject to the restrictions in section 70(2) and (3).

(2) The arbitral tribunal may continue the arbitral proceedings and make a further award while an application to the court under this section is pending in relation to an award as to jurisdiction.

(3) On an application under this section challenging an award of the arbitral tribunal as to its substantive jurisdiction, the court may by order —
  (a) confirm the award,
  (b) vary the award, or
  (c) set aside the award in whole or in part.

(4) The leave of the court is required for any appeal from a decision of the court under this section.

**68.** — (1) A party to arbitral proceedings may (upon notice to the other parties and to the tribunal) apply to the court challenging an award in the proceedings on the ground of serious irregularity affecting the tribunal, the proceedings or the award.

A party may lose the right to object (see section 73) and the right to apply is subject to the restrictions in section 70(2) and (3).

Challenging the award: serious irregularity.

(2) Serious irregularity means an irregularity of one or more of the following kinds which the court considers has caused or will cause substantial injustice to the applicant —

(a) failure by the tribunal to comply with section 33 (general duty of tribunal);

(b) the tribunal exceeding its powers (otherwise than by exceeding its substantive jurisdiction: see section 67);

(c) failure by the tribunal to conduct the proceedings in accordance with the procedure agreed by the parties;

(d) failure by the tribunal to deal with all the issues that were put to it;

(e) any arbitral or other institution or person vested by the parties with powers in relation to the proceedings or the award exceeding its powers;

(f) uncertainty or ambiguity as to the effect of the award;

(g) the award being obtained by fraud or the award or the way in which it was procured being contrary to public policy;

(h) failure to comply with the requirements as to the form of the award; or

(i) any irregularity in the conduct of the proceedings or in the award which is admitted by the tribunal or by any arbitral or other institution or person vested by the parties with powers in relation to the proceedings or the award.

(3) If there is shown to be serious irregularity affecting the tribunal, the proceedings or the award, the court may —

(a) remit the award to the tribunal, in whole or in part, for reconsideration,

(b) set the award aside in whole or in part, or

(c) declare the award to be of no effect, in whole or in part.

The court shall not exercise its power to set aside or to declare an award to be of no effect, in whole or in part, unless it is satisfied that it would be inappropriate to remit the matters in question to the tribunal for reconsideration.

(4) The leave of the court is required for any appeal from a decision of the court under this section.

**69.** — (1) Unless otherwise agreed by the parties, a party to arbitral proceedings may (upon notice to the other parties and to the tribunal) appeal to the court on a question of law arising out of an award made in the proceedings.

Appeal on point of law.

An agreement to dispense with reasons for the tribunal's award shall be considered an agreement to exclude the court's jurisdiction under this section.

(2) An appeal shall not be brought under this section except—
  (a) with the agreement of all the other parties to the proceedings, or
  (b) with the leave of the court.
The right to appeal is also subject to the restrictions in section 70(2) and (3).

(3) Leave to appeal shall be given only if the court is satisfied—
  (a) that the determination of the question will substantially affect the rights of one or more of the parties,
  (b) that the question is one which the tribunal was asked to determine,
  (c) that, on the basis of the findings of fact in the award—
    (i) the decision of the tribunal on the question is obviously wrong, or
    (ii) the question is one of general public importance and the decision of the tribunal is at least open to serious doubt, and
  (d) that, despite the agreement of the parties to resolve the matter by arbitration, it is just and proper in all the circumstances for the court to determine the question.

(4) An application for leave to appeal under this section shall identify the question of law to be determined and state the grounds on which it is alleged that leave to appeal should be granted.

(5) The court shall determine an application for leave to appeal under this section without a hearing unless it appears to the court that a hearing is required.

(6) The leave of the court is required for any appeal from a decision of the court under this section to grant or refuse leave to appeal.

(7) On an appeal under this section the court may by order—
  (a) confirm the award,
  (b) vary the award,
  (c) remit the award to the tribunal, in whole or in part, for reconsideration in the light of the court's determination, or
  (d) set aside the award in whole or in part.
The court shall not exercise its power to set aside an award, in whole or in part, unless it is satisfied that it would be inappropriate to remit the matters in question to the tribunal for reconsideration.

(8) The decision of the court on an appeal under this section shall be treated as a judgment of the court for the purposes of a further appeal.
But no such appeal lies without the leave of the court which shall not be given unless the court considers that the question is one of general importance or is one which for some other special reason should be considered by the Court of Appeal.

**70.** — (1) The following provisions apply to an application or appeal under section 67, 68 or 69.

Challenge or appeal: supplementary provisions.

(2) An application or appeal may not be brought if the applicant or appellant has not first exhausted —

(a) any available arbitral process of appeal or review, and

(b) any available recourse under section 57 (correction of award or additional award).

(3) Any application or appeal must be brought within 28 days of the date of the award or, if there has been any arbitral process of appeal or review, of the date when the applicant or appellant was notified of the result of that process.

(4) If on an application or appeal it appears to the court that the award —

(a) does not contain the tribunal's reasons, or

(b) does not set out the tribunal's reasons in sufficient detail to enable the court properly to consider the application or appeal,

the court may order the tribunal to state the reasons for its award in sufficient detail for that purpose.

(5) Where the court makes an order under subsection (4), it may make such further order as it thinks fit with respect to any additional costs of the arbitration resulting from its order.

(6) The court may order the applicant or appellant to provide security for the costs of the application or appeal, and may direct that the application or appeal be dismissed if the order is not complied with.

The power to order security for costs shall not be exercised on the ground that the applicant or appellant is —

(a) an individual ordinarily resident outside the United Kingdom, or

(b) a corporation or association incorporated or formed under the law of a country outside the United Kingdom, or whose central management and control is exercised outside the United Kingdom.

(7) The court may order that any money payable under the award shall be brought into court or otherwise secured pending the determination of the application or appeal, and may direct that the application or appeal be dismissed if the order is not complied with.

(8) The court may grant leave to appeal subject to conditions to the same or similar effect as an order under subsection (6) or (7).

This does not affect the general discretion of the court to grant leave subject to conditions.

**71.** — (1) The following provisions have effect where the court makes an order under section 67, 68 or 69 with respect to an award.

Challenge or appeal: effect of order of court.

(2) Where the award is varied, the variation has effect as part of the tribunal's award.

(3) Where the award is remitted to the tribunal, in whole or in part, for reconsideration, the tribunal shall make a fresh award in respect of the matters remitted within three months of the date of the order for remission or such longer or shorter period as the court may direct.

(4) Where the award is set aside or declared to be of no effect, in whole or in part, the court may also order that any provision that an award is a condition precedent to the bringing of legal proceedings in respect of a matter to which the arbitration agreement applies, is of no effect as regards the subject matter of the award or, as the case may be, the relevant part of the award.

*Miscellaneous*

**Saving for rights of person who takes no part in proceedings.**

**72.** — (1) A person alleged to be a party to arbitral proceedings but who takes no part in the proceedings may question —
  (a) whether there is a valid arbitration agreement,
  (b) whether the tribunal is properly constituted, or
  (c) what matters have been submitted to arbitration in accordance with the arbitration agreement,

by proceedings in the court for a declaration or injunction or other appropriate relief.

(2) He also has the same right as a party to the arbitral proceedings to challenge an award —
  (a) by an application under section 67 on the ground of lack of substantive jurisdiction in relation to him, or
  (b) by an application under section 68 on the ground of serious irregularity (within the meaning of that section) affecting him;

and section 70(2) (duty to exhaust arbitral procedures) does not apply in his case.

**Loss of right to object.**

**73.** — (1) If a party to arbitral proceedings takes part, or continues to take part, in the proceedings without making, either forthwith or within such time as is allowed by the arbitration agreement or the tribunal or by any provision of this Part, any objection —
  (a) that the tribunal lacks substantive jurisdiction,
  (b) that the proceedings have been improperly conducted,
  (c) that there has been a failure to comply with the arbitration agreement or with any provision of this Part, or
  (d) that there has been any other irregularity affecting the tribunal or the proceedings,

he may not raise that objection later, before the tribunal or the court, unless he shows that, at the time he took part or continued to take part in the proceedings, he did not know and could not with reasonable diligence have discovered the grounds for the objection.

(2) Where the arbitral tribunal rules that it has substantive jurisdiction and a party to arbitral proceedings who could have questioned that ruling —

(a) by any available arbitral process of appeal or review, or

(b) by challenging the award,

does not do so, or does not do so within the time allowed by the arbitration agreement or any provision of this Part, he may not object later to the tribunal's substantive jurisdiction on any ground which was the subject of that ruling.

**74.**—(1) An arbitral or other institution or person designated or requested by the parties to appoint or nominate an arbitrator is not liable for anything done or omitted in the discharge or purported discharge of that function unless the act or omission is shown to have been in bad faith.

(2) An arbitral or other institution or person by whom an arbitrator is appointed or nominated is not liable, by reason of having appointed or nominated him, for anything done or omitted by the arbitrator (or his employees or agents) in the discharge or purported discharge of his functions as arbitrator.

(3) The above provisions apply to an employee or agent of an arbitral or other institution or person as they apply to the institution or person himself.

*Immunity of arbitral institutions, &c.*

**75.** The powers of the court to make declarations and orders under section 73 of the Solicitors Act 1974 or Article 71H of the Solicitors (Northern Ireland) Order 1976 (power to charge property recovered in the proceedings with the payment of solicitors' costs) may be exercised in relation to arbitral proceedings as if those proceedings were proceedings in the court.

*Charge to secure payment of solicitors' costs.*

### Supplementary

**76.**—(1) The parties are free to agree on the manner of service of any notice or other document required or authorised to be given or served in pursuance of the arbitration agreement or for the purposes of the arbitral proceedings.

*Service of notices, &c.*

(2) If or to the extent that there is no such agreement the following provisions apply.

(3) A notice or other document may be served on a person by any effective means.

(4) If a notice or other document is addressed, pre-paid and delivered by post—

(a) to the addressee's last known principal residence or, if he is or has been carrying on a trade, profession or business, his last known principal business address, or

(b) where the addressee is a body corporate, to the body's registered or principal office,

it shall be treated as effectively served.

(5) This section does not apply to the service of documents for the purposes of legal proceedings, for which provision is made by rules of court.

(6) References in this Part to a notice or other document include any form of communication in writing and references to giving or serving a notice or other document shall be construed accordingly.

**Powers of court in relation to service of documents.**
77.—(1) This section applies where service of a document on a person in the manner agreed by the parties, or in accordance with provisions of section 76 having effect in default of agreement, is not reasonably practicable.

(2) Unless otherwise agreed by the parties, the court may make such order as it thinks fit—
    (a) for service in such manner as the court may direct, or
    (b) dispensing with service of the document.

(3) Any party to the arbitration agreement may apply for an order, but only after exhausting any available arbitral process for resolving the matter.

(4) The leave of the court is required for any appeal from a decision of the court under this section.

**Reckoning periods of time.**
78.—(1) The parties are free to agree on the method of reckoning periods of time for the purposes of any provision agreed by them or any provision of this Part having effect in default of such agreement.

(2) If or to the extent there is no such agreement, periods of time shall be reckoned in accordance with the following provisions.

(3) Where the act is required to be done within a specified period after or from a specified date, the period begins immediately after that date.

(4) Where the act is required to be done a specified number of clear days after a specified date, at least that number of days must intervene between the day on which the act is done and that date.

(5) Where the period is a period of seven days or less which would include a Saturday, Sunday or a public holiday in the place where anything which has to be done within the period falls to be done, that day shall be excluded.
    In relation to England and Wales or Northern Ireland, a "public holiday" means Christmas Day, Good Friday or a day which under the Banking and Financial Dealings Act 1971 is a bank holiday.

**Power of court to extend time limits relating to arbitral proceedings.**
79.—(1) Unless the parties otherwise agree, the court may by order extend any time limit agreed by them in relation to any matter relating to the arbitral proceedings or specified in any provision of this Part having effect in default of such agreement.

This section does not apply to a time limit to which section 12 applies (power of court to extend time for beginning arbitral proceedings, &c.).

(2) An application for an order may be made—
  (a) by any party to the arbitral proceedings (upon notice to the other parties and to the tribunal), or
  (b) by the arbitral tribunal (upon notice to the parties).

(3) The court shall not exercise its power to extend a time limit unless it is satisfied—
  (a) that any available recourse to the tribunal, or to any arbitral or other institution or person vested by the parties with power in that regard, has first been exhausted, and
  (b) that a substantial injustice would otherwise be done.

(4) The court's power under this section may be exercised whether or not the time has already expired.

(5) An order under this section may be made on such terms as the court thinks fit.

(6) The leave of the court is required for any appeal from a decision of the court under this section.

**80.**—(1) References in this Part to an application, appeal or other step in relation to legal preceedings being taken "upon notice" to the other parties to the arbitral proceedings, or to the tribunal, are to such notice of the originating process as is required by rules of court and do not impose any separate requirement.

*Notice and other requirements in connection with legal proceedings.*

(2) Rules of court shall be made—
  (a) requiring such notice to be given as indicated by any provision of this Part, and
  (b) as to the manner, form and content of any such notice.

(3) Subject to any provision made by rules of court, a requirement to give notice to the tribunal of legal proceedings shall be construed—
  (a) if there is more than one arbitrator, as a requirement to give notice to each of them; and
  (b) if the tribunal is not fully constituted, as a requirement to give notice to any arbitrator who has been appointed.

(4) References in this Part to making an application or appeal to the court within a specified period are to the issue within that period of the appropriate originating process in accordance with rules of court.

(5) Where any provision of this Part requires an application or appeal to be made to the court within a specified time, the rules of court relating to the reckoning of periods, the extending or abridging of periods, and the consequences of not taking a step within the period prescribed by the rules, apply in relation to that requirement.

(6) Provision may be made by rules of court amending the provisions of this Part —

    (a) with respect to the time within which any application or appeal to the court must be made,

    (b) so as to keep any provision made by this Part in relation to arbitral proceedings in step with the corresponding provision of rules of court applying in relation to proceedings in the court, or

    (c) so as to keep any provision made by this Part in relation to legal proceedings in step with the corresponding provision of rules of court applying generally in relation to proceedings in the court.

(7) Nothing in this section affects the generality of the power to make rules of court.

**Saving for certain matters governed by common law.**

81. — (1) Nothing in this Part shall be construed as excluding the operation of any rule of law consistent with the provisions of this Part, in particular, any rule of law as to —

    (a) matters which are not capable of settlement by arbitration;

    (b) the effect of an oral arbitration agreement; or

    (c) the refusal of recognition or enforcement of an arbitral award on grounds of public policy.

(2) Nothing in this Act shall be construed as reviving any jurisdiction of the court to set aside or remit an award on the ground of errors of fact or law on the face of the award.

**Minor definitions.**

82. — (1) In this Part —

"arbitrator", unless the context otherwise requires, includes an umpire;

"available arbitral process", in relation to any matter, includes any process of appeal to or review by an arbitral or other institution or person vested by the parties with powers in relation to that matter;

"claimant", unless the context otherwise requires, includes a counter-claimant, and related expressions shall be construed accordingly;

"dispute" includes any difference;

"enactment" includes an enactment contained in Northern Ireland legislation;

"legal proceedings" means civil proceedings in the High Court or a county court;

"peremptory order" means an order made under section 41(5) or made in exercise of any corresponding power conferred by the parties;

"premises" includes land, buildings, moveable structures, vehicles, vessels, aircraft and hovercraft;

"question of law" means —

    (a) for a court in England and Wales, a question of the law of England and Wales, and

    (b) for a court in Northern Ireland, a question of the law of Northern Ireland;

"substantive jurisdiction", in relation to an arbitral tribunal, refers to the matters specified in section 30(1)(a) to (c), and references to the tribunal exceeding its substantive jurisdiction shall be construed accordingly.

(2) References in this Part to a party to an arbitration agreement include any person claiming under or through a party to the agreement.

**83.** In this Part the expressions listed below are defined or otherwise explained by the provisions indicated —

*Index of defined expressions: Part I*

| | |
|---|---|
| agreement, agree and agreed | section 5(1) |
| agreement in writing | section 5(2) to (5) |
| arbitration agreement | sections 6 and 5(1) |
| arbitrator | section 82(1) |
| available arbitral process | section 82(1) |
| claimant | section 82(1) |
| commencement (in relation to arbitral proceedings) | section 14 |
| costs of the arbitration | section 59 |
| the court | section 105 |
| dispute | section 82(1) |
| enactment | section 82(1) |
| legal proceedings | section 82(1) |
| Limitation Acts | section 13(4) |
| notice (or other document) | section 76(6) |
| party — | |
|   — in relation to an arbitration agreement | section 82(2) |
|   — where section 106(2) or (3) applies | section 106(4) |
| peremptory order | section 82(1) (and see section 41(5)) |
| premises | section 82(1) |
| question of law | section 82(1) |
| recoverable costs | sections 63 and 64 |
| seat of the arbitration | section 3 |
| serve and service (of notice or other document) | section 76(6) |
| substantive jurisdiction (in relation to an arbitral tribunal) | section 82(1) (and see section 30(1)(a) to (c)) |
| upon notice (to the parties or the tribunal) | section 80 |
| written and in writing | section 5(6) |

**84.** — (1) The provisions of this Part do not apply to arbitral proceedings commenced before the date on which this Part comes into force.

*Transitional provisions.*

(2) They apply to arbitral proceedings commenced on or after that date under an arbitration agreement whenever made.

(3) The above provisions have effect subject to any transitional provision made by an order under section 109(2) (power to include transitional provisions in commencement order).

## PART II

OTHER PROVISIONS RELATING TO ARBITRATION

### *Domestic arbitration agreements*

**Modification of Part I in relation to domestic arbitration agreement.**

**85.** — (1) In the case of a domestic arbitration agreement the provisions of Part I are modified in accordance with the following sections.

(2) For this purpose a "domestic arbitration agreement" means an arbitration agreement to which none of the parties is —
  (a) an individual who is a national of, or habitually resident in, a state other than the United Kingdom, or
  (b) a body corporate which is incorporated in, or whose central control and management is exercised in, a state other than the United Kingdom,
and under which the seat of the arbitration (if the seat has been designated or determined) is in the United Kingdom.

(3) In subsection (2) "arbitration agreement" and "seat of the arbitration" have the same meaning as in Part I (see sections 3, 5(1) and 6).

**Staying of legal proceedings.**

**86.** — (1) In section 9 (stay of legal proceedings), subsection (4) (stay unless the arbitration agreement is null and void, inoperative, or incapable of being performed) does not apply to a domestic arbitration agreement.

(2) On an application under that section in relation to a domestic arbitration agreement the court shall grant a stay unless satisfied —
  (a) that the arbitration agreement is null and void, inoperative, or incapable of being performed, or
  (b) that there are other sufficient grounds for not requiring the parties to abide by the arbitration agreement.

(3) The court may treat as a sufficient ground under subsection (2)(b) the fact that the applicant is or was at any material time not ready and willing to do all things necessary for the proper conduct of the arbitration or of any other dispute resolution procedures required to be exhausted before resorting to arbitration.

(4) For the puposes of this section the question whether an arbitration agreement is a domestic arbitration agreement shall be determined by reference to the facts at the time the legal proceedings are commenced.

87.—(1) In the case of a domestic arbitration agreement any agreement to exclude the jurisdiction of the court under—
    (a) section 45 (determination of preliminary point of law), or
    (b) section 69 (challenging the award: appeal on point of law),
is not effective unless entered into after the commencement of the arbitral proceedings in which the question arises or the award is made.

    *Effectiveness of agreement to exclude court's jurisdiction.*

(2) For this purpose the commencement of the arbitral proceedings has the same meaning as in Part I (see section 14).

(3) For the purposes of this section the question whether an arbitration agreement is a domestic arbitration agreement shall be determined by reference to the facts at the time the agreement is entered into.

88.—(1) The Secretary of State may by order repeal or amend the provisions of sections 85 to 87.

    *Power to repeal or amend sections 85 to 87.*

(2) An order under this section may contain such supplementary, incidental and transitional provisions as appear to the Secretary of State to be appropriate.

(3) An order under this section shall be made by statutory instrument and no such order shall be made unless a draft of it has been laid before and approved by a resolution of each House of Parliament.

### Consumer arbitration agreements

89.—(1) The following sections extend the application of the Unfair Terms in Consumer Contracts Regulations 1994 in relation to a term which constitutes an arbitration agreement.
    For this purpose "arbitration agreement" means an agreement to submit to arbitration present or future disputes or differences (whether or not contractual).

    *Application of unfair terms regulations to consumer arbitration agreements.*

(2) In those sections "the Regulations" means those regulations and includes any regulations amending or replacing those regulations.

(3) Those sections apply whatever the law applicable to the arbitration agreement.

90. The Regulations apply where the consumer is a legal person as they apply where the consumer is a natural person.

    *Regulations apply where consumer is a legal person.*

91.—(1) A term which constitutes an arbitration agreement is unfair for the purposes of the Regulations so far as it relates to a claim for a pecuniary remedy which does not exceed the amount specified by order for the purposes of this section.

    *Arbitration agreement unfair where modest amount sought.*

(2) Orders under this section may make different provision for different cases and for different purposes.

(3) The power to make orders under this section is exercisable —
  (a) for England and Wales, by the Secretary of State with the concurrence of the Lord Chancellor,
  (b) for Scotland, by the Secretary of State with the concurrence of the Lord Advocate, and
  (c) for Northern Ireland, by the Department of Economic Development for Northern Ireland with the concurrence of the Lord Chancellor.

(4) Any such order for England and Wales or Scotland shall be made by statutory instrument which shall be subject to annulment in pursuance of a resolution of either House of Parliament.

(5) Any such order for Northern Ireland shall be a statutory rule for the purposes of the Statutory Rules (Northern Ireland) Order 1979 and shall be subject to negative resolution, within the meaning of section 41(6) of the Interpretation Act (Northern Ireland) 1954.

### Small claims arbitration in the county court

**Exclusion of Part I in relation to small claims arbitration in the county court.**
**92.** Nothing in Part I of this Act applies to arbitration under section 64 of the County Courts Act 1984.

### Appointment of judges as arbitrators

**Appointment of judges as arbitrators.**
**93.** — (1) A judge of the Commercial Court or an official referee may, if in all the circumstances he thinks fit, accept appointment as a sole arbitrator or as umpire by or by virtue of an arbitration agreement.

(2) A judge of the Commercial Court shall not do so unless the Lord Chief Justice has informed him that, having regard to the state of business in the High Court and the Crown Court, he can be made available.

(3) An official referee shall not do so unless the Lord Chief Justice has informed his that, having regard to the state of official referees' business, he can be made available.

(4) The fees payable for the services of a judge of the Commercial Court or official referee as arbitrator or umpire shall be taken in the High Court.

(5) In this section —
  "arbitration agreement" has the same meaning as in Part I; and
  "official referee" means a person nominated under section 68(1)(a) of the Supreme Court Act 1981 to deal with official referees' business.

(6) The provisions of Part I of this Act apply to arbitration before a person appointed under this section with the modifications specified in Schedule 2.

*Statutory arbitrations*

**94.**—(1) The provisions of Part I apply to every arbitration under an enactment (a "statutory arbitration"), whether the enactment was passed or made before or after the commencement of this Act, subject to the adaptations and exclusions specified in sections 95 to 98.

(2) The provisions of Part I do not apply to a statutory arbitration if or to the extent that their application—

    (a) is inconsistent with the provisions of the enactment concerned, with any rules or procedure authorised or recognised by it, or

    (b) is excluded by any other enactment.

(3) In this section and the following provisions of this Part "enactment"—

    (a) in England and Wales, includes an enactment contained in subordinate legislation within the meaning of the Interpretation Act 1978;

    (b) in Northern Ireland, means a statutory provision within the meaning of section 1(f) of the Interpretation Act (Northern Ireland) 1954.

*Application of Part I to statutory arbitrations.*

**95.**—(1) The provisions of Part I apply to a statutory arbitration—

    (a) as if the arbitration were pursuant to an arbitration agreement and as if the enactment were that agreement, and

    (b) as if the persons by and against whom a claim subject to arbitration in pursuance of the enactment may be or has been made were parties to that agreement.

(2) Every statutory arbitration shall be taken to have its seat in England and Wales or, as the case may be, in Northern Ireland.

*General adaptation of provisions in relation to statutory arbitrations.*

**96.**—(1) The following provisions of Part I apply to a statutory arbitration with the following adaptations.

(2) In section 30(1) (competence of tribunal to rule on its own jurisdiction), the reference in paragraph (a) to whether there is a valid arbitration agreement shall be construed as a reference to whether the enactment applies to the dispute or difference in question.

(3) Section 35 (consolidation of proceedings and concurrent hearings) applies only so as to authorise the consolidation of proceedings, or concurrent hearings in proceedings, under the same enactment.

(4) Section 46 (rules applicable to substance of dispute) applies with the omission of subsection (1)(b) (determination in accordance with considerations agreed by parties).

*Specific adaptations of provisions in relation to statutory arbitrations.*

Provisions
excluded from
applying to
statutory
arbitrations.

**97.** The following provisions of Part I do not apply in relation to a statutory arbitration—

(a) section 8 (whether agreement discharged by death of a party);

(b) section 12 (power of court to extend agreed time limits);

(c) sections 9(5), 10(2) and 71(4) (restrictions on effect of provision that award condition precedent to right to bring legal proceedings).

Power to make
further provision
by regulations.

**98.**—(1) The Secretary of State may make provision by regulations for adapting or excluding any provision of Part I in relation to statutory arbitrations in general or statutory arbitrations of any particular description.

(2) The power is exercisable whether the enactment concerned is passed or made before or after the commencement of this Act.

(3) Regulations under this section shall be made by statutory instrument which shall be subject to annulment in pursuance of a resolution of either House of Parliament.

PART III

RECOGNITION AND ENFORCEMENT OF CERTAIN FOREIGN AWARDS

*Enforcement of Geneva Convention awards*

Continuation of
Part II of the
Arbitration Act
1950.

**99.** Part II of the Arbitration Act 1950 (enforcement of certain foreign awards) continues to apply in relation to foreign awards within the meaning of that Part which are not also New York Convention awards.

*Recognition and enforcement of New York Convention awards*

New York
Convention
awards.

**100.**—(1) In this Part a "New York Convention award" means an award made, in pursuance of an arbitration agreement, in the territory of a state (other than the United Kingdom) which is a party to the New York Convention.

(2) For the purposes of subsection (1) and of the provisions of this Part relating to such awards—

(a) "arbitration agreement" means an arbitration agreement in writing, and

(b) an award shall be treated as made at the seat of the arbitration, regardless of where it was signed, despatched or delivered to any of the parties.

In this subsection "agreement in writing" and "seat of the arbitration" have the same meaning as in Part I.

(3) If Her Majesty by Order in Council declares that a state specified in the Order is a party to the New York Convention, or is a party in respect of any territory so specified, the Order shall, while in force, be conclusive evidence of that fact.

(4) In this section "the New York Convention" means the Convention on the Recognition and Enforcement of Foreign Arbitral Awards adopted

by the United Nations Conference on International Commercial Arbitration on 10th June 1958.

**101.** — (1) A New York Convention award shall be recognised as binding on the persons as between whom it was made, and may accordingly be relied on by those persons by way of defence, set-off or otherwise in any legal proceedings in England and Wales or Northern Ireland.

*Recognition and enforcement of awards.*

(2) A New York Convention award may, by leave of the court, be enforced in the same manner as a judgment or order of the court to the same effect.
As to the meaning of "the court" see section 105.

(3) Where leave is so given, judgment may be entered in terms of the award.

**102.** — (1) A party seeking the recognition or enforcement of a New York Convention award must produce —

*Evidence to be produced by party seeking recognition or enforcement.*

    (a) the duly authenticated original award or a duly certified copy of it, and
    (b) the original arbitration agreement or a duly certified copy of it.

(2) If the award or agreement is in a foreign language, the party must also produce a translation of it certified by an official or sworn translator or by a diplomatic or consular agent.

**103.** — (1) Recognition or enforcement of a New York Convention award shall not be refused except in the following cases.

*Refusal of recognition or enforcement.*

(2) Recognition or enforcement of the award may be refused if the person against whom it is invoked proves —
    (a) that a party to the arbitration agreement was (under the law applicable to him) under some incapacity;
    (b) that the arbitration agreement was not valid under the law to which the parties subjected it or, failing any indication thereon, under the law of the country where the award was made;
    (c) that he was not given proper notice of the appointment of the arbitrator or of the arbitration proceedings or was otherwise unable to present his case;
    (d) that the award deals with a difference not contemplated by or not falling within the terms of the submission to arbitration or contains decisions on matters beyond the scope of the submission to arbitration (but see subsection (4));
    (e) that the composition of the arbitral tribunal or the arbitral procedure was not in accordance with the agreement of the parties or, failing such agreement, with the law of the country in which the arbitration took place;
    (f) that the award has not yet become binding on the parties, or has

been set aside or suspended by a competent authority of the country in which, or under the law of which, it was made.

(3) Recognition or enforcement of the award may also be refused if the award as in respect of a matter which is not capable of settlement by arbitration, or if it would be contrary to public policy to recognise or enforce the award.

(4) An award which contains decisions on matters not submitted to arbitration may be recognised or enforced to the extent that it contains decisions on matters submitted to arbitration which can be separated from those on matters not so submitted.

(5) Where an application for the setting aside or suspension of the award has been made to such a competent authority as is mentioned in subsection (2)(f), the court before which the award is sought to be relied upon may, if it considers it proper, adjourn the decision on the recognition or enforcement of the award.

It may also on the application of the party claiming recognition or enforcement of the award order the other party to give suitable security.

**Saving for other bases of recognition or enforcement.**

**104.** Nothing in the preceding provisions of this Part affects any right to rely upon or enforce a New York Convention award at common law or under section 66.

PART IV

GENERAL PROVISIONS

**Meaning of "the court": jurisdiction of High Court and county court.**

**105.** — (1) In this Act "the court" means the High Court or a county court, subject to the following provisions.

(2) The Lord Chancellor may by order make provision —
  (a) allocating proceedings under this Act to the High Court or to county courts; or
  (b) specifying proceedings under this Act which may be commenced or taken only in the High Court or in a county court.

(3) The Lord Chancellor may by order make provision requiring proceedings of any specified description under this Act in relation to which a county court has jurisdiction to be commenced or taken in one or more specified county courts.

Any jurisdiction so exercisable by a specified county court is exercisable throughout England and Wales or, as the case may be, Northern Ireland.

(4) An order under this section —
  (a) may differentiate between categories of proceedings by reference to such criteria as the Lord Chancellor sees fit to specify, and
  (b) may make such incidental or transitional provision as the Lord Chancellor considers necessary or expedient.

(5) An order under this section for England and Wales shall be made by

statutory instrument which shall be subject to annulment in pursuance of a resolution of either House of Parliament.

(6) An order under this section for Northern Ireland shall be a statutory rule for the purposes of the Statutory Rules (Northern Ireland) Order 1979 which shall be subject to annulment in pursuance of a resolution of either House of Parliament in like manner as a statutory instrument and section 5 of the Statutory Instruments Act 1946 shall apply accordingly.

**106.**—(1) Part I of this Act applies to any arbitration agreement to which Her Majesty, either in right of the Crown or of the Duchy of Lancaster or otherwise, or the Duke of Cornwall, is a party. Crown application.

(2) Where Her Majesty is party to an arbitration agreement otherwise than in right of the Crown, Her Majesty shall be represented for the purposes of any arbitral proceedings—
- (a) where the agreement was entered into by Her Majesty in right of the Duchy of Lancaster, by the Chancellor of the Duchy or such person as he may appoint, and
- (b) in any other case, by such person as Her Majesty may appoint in writing under the Royal Sign Manual.

(3) Where the Duke of Cornwall is party to an arbitration agreement, he shall be represented for the purposes of any arbitral proceedings by such person as he may appoint.

(4) References in Part I to a party or the parties to the arbitration agreement or to arbitral proceedings shall be construed, where subsection (2) or (3) applies, as references to the person representing Her Majesty or the Duke of Cornwall.

**107.**—(1) The enactments specified in Schedule 3 are amended in accordance with that Schedule, the amendments being consequential on the provisions of this Act. Consequential amendments and repeals.

(2) The enactments specified in Schedule 4 are repealed to the extent specified.

**108.**—(1) The provisions of this Act extend to England and Wales and, except as mentioned below, to Northern Ireland. Extent.

(2) The following provisions of Part II do not extend to Northern Ireland—

section 92 (exclusion of Part I in relation to small claims arbitration in the county court), and

section 93 and Schedule 2 (appointment of judges as arbitrators).

(3) Sections 89, 90 and 91 (consumer arbitration agreements) extend to Scotland and the provisions of Schedules 3 and 4 (consequential amend-

231

ments and repeals) extend to Scotland so far as they relate to enactments which so extend, subject as follows.

(4) The repeal of the Arbitration Act 1975 extends only to England and Wales and Northern Ireland.

Commencement.     **109.**−(1) The provisions of this Act come into force on such day as the Secretary of State may appoint by order made by statutory instrument, and different days may be appointed for different purposes.

(2) An order under subsection (1) may contain such transitional provisions as appear to the Secretary of State to be appropriate.

Short title.     **110.** This Act may be cited as the Arbitration Act 1996.

# SCHEDULES

## SCHEDULE 1

### MANDATORY PROVISIONS OF PART I

sections 9 to 11 (stay of legal proceedings);

section 12 (power of court to extend agreed time limits);

section 13 (application of Limitation Acts);

section 24 (power of court to remove arbitrator);

section 26(1) (effect of death of arbitrator);

section 28 (liability of parties for fees and expenses of arbitrators);

section 29 (immunity of arbitrator);

section 31 (objection to substantive jurisdiction of tribunal);

section 32 (determination of preliminary point of jurisdiction);

section 33 (general duty of tribunal);

section 37(2) (items to be treated as expenses of arbitrators);

section 40 (general duty of parties);

section 43 (securing the attendance of witnesses);

section 56 (power to withhold award in case of non-payment);

section 60 (effectiveness of agreement for payment of costs in any event);

section 66 (enforcement of award);

sections 67 and 68 (challenging the award: substantive jurisdiction and serious irregularity), and sections 70 and 71 (supplementary provisions; effect of order of court) so far as relating to those sections;

section 72 (saving for rights of person who takes no part in proceedings);

section 73 (loss of right to object);

section 74 (immunity of arbitral institutions, &c.);

section 75 (charge to secure payment of solicitors' costs).

# APPENDIX C
# THE ARBITRATION ACT 1950:
# PART II

ENFORCEMENT OF CERTAIN FOREIGN AWARDS

**35.**—(1) This Part of this Act applies to any award made after the twenty-eighth day of July, nineteen hundred and twenty-four—

*Awards to which Part II applies.*

(a) in pursuance of an agreement for arbitration to which the protocol set out in the First Schedule to this Act applies; and

(b) between persons of whom one is subject to the jurisdiction of some one of such Powers as His Majesty, being satisfied that reciprocal provisions have been made, may by Order in Council declare to be parties to the convention set out in the Second Schedule to this Act, and of whom the other is subject to the jurisdiction of some other of the Powers aforesaid; and

(c) in one of such territories as His Majesty, being satisfied that reciprocal provisions have been made, may by Order in Council declare to be territories to which the said convention applies;

and an award to which this Part of this Act applies is in this Part of this Act referred to as "a foreign award".

(2) His Majesty may by a subsequent Order in Council vary or revoke any Order previously made under this section.

(3) Any Order in Council under section one of the Arbitration (Foreign Awards) Act, 1930, which is in force at the commencement of this Act shall have effect as if it had been made under this section.

**36.**—(1) A foreign award shall, subject to the provisions of this Part of this Act, be enforceable in England either by action or in the same manner as the award of an arbitrator is enforceable by virtue of section twenty-six of this Act.

*Effect of foreign awards.*

(2) Any foreign award which would be enforceable under this Part of this Act shall be treated as binding for all purposes on the persons as between whom it was made, and may accordingly be relied on by any of those persons by way of defence, set off or otherwise in any legal proceedings in England, and any references in this Part of this Act to enforcing a foreign award shall be construed as including references to relying on an award.

**37.**—(1) In order that a foreign award may be enforceable under this Part of this Act it must have—

 (*a*) been made in pursuance of an agreement for arbitration which was valid under the law by which it was governed;

 (*b*) been made by the tribunal provided for in the agreement or constituted in manner agreed upon by the parties;

 (*c*) been made in conformity with the law governing the arbitration procedure;

 (*d*) become final in the country in which it was made;

 (*e*) been in respect of a matter which may lawfully be referred to arbitration under the law of England;

and the enforcement thereof must not be contrary to the public policy or the law of England.

(2) Subject to the provisions of this subsection, a foreign award shall not be enforceable under this Part of this Act if the court dealing with the case is satisfied that—

 (*a*) the award has been annulled in the country in which it was made; or

 (*b*) the party against whom it is sought to enforce the award was not given notice of the arbitration proceedings in sufficient time to enable him to present his case, or was under some legal incapacity and was not properly represented; or

 (*c*) the award does not deal with all the questions referred or contains decisions on matters beyond the scope of the agreement for arbitration:

Provided that, if the award does not deal with all the questions referred, the court may, if it thinks fit, either postpone the enforcement of the award or order its enforcement subject to the giving of such security by the person seeking to enforce it as the court may think fit.

(3) If a party seeking to resist the enforcement of a foreign award proves that there is any ground other than the non-existence of the conditions specified in paragraphs (a), (b) and (c) of subsection (1) of this section, or the existence of the conditions specified in paragraphs (b) and (c) of subsection (2) of this section, entitling him to contest the validity of the award, the court may, if it thinks fit, either refuse to enforce the award or adjourn the hearing until after the expiration of such period as appears to the court to be reasonably sufficient to enable that party to take the necessary steps to have the award annulled by the competent tribunal.

**38.**—(1) The party seeking to enforce a foreign award must produce—

 (*a*) the original award or a copy thereof duly authenticated in manner required by the law of the country in which it was made; and

 (*b*) evidence proving that the award has become final; and

 (*c*) such evidence as may be necessary to prove that the award is a foreign award and that the conditions mentioned in paragraphs (a), (b) and (c) of subsection (1) of the last foregoing section are satisfied.

(2) In any case where any document required to be produced under subsection (1) of this section is in a foreign language, it shall be the duty of the party seeking to enforce the award to produce a translation certified as correct by a diplomatic or consular agent of the country to which that party belongs, or certified as correct in such other manner as may be sufficient according to the law of England.

(3) Subject to the provisions of this section, rules of court may be made under section ninety-nine of the Supreme Court of Judicature (Consolidation) Act, 1925, with respect to the evidence which must be furnished by a party seeking to enforce an award under this Part of this Act.

**39.** For the purposes of this Part of this Act, an award shall not be deemed final if any proceedings for the purpose of contesting the validity of the award are pending in the country in which it was made. *Meaning of "final award".*

**40.** Nothing in this Part of this Act shall—

(a) prejudice any rights which any person would have had of enforcing in England any award or of availing himself in England of any award if neither this Part of this Act nor Part I of the Arbitration (Foreign Awards) Act, 1930, had been enacted; or

(b) apply to any award made on an arbitration agreement governed by the law of England. *Saving for other rights &c.*

**41.**—(1) The following provisions of this section shall have effect for the purpose of the application of this Part of this Act to Scotland. *Application of Part II to Scotland.*

(2) For the references to England there shall be substituted references to Scotland.

(3) For subsection (1) of section thirty-six there shall be substituted the following subsection:—

"(1) A foreign award shall, subject to the provisions of this Part of this Act, be enforceable by action, or, if the agreement for arbitration contains consent to the registration of the award in the Books of Council and Session for execution and the award is so registered, it shall, subject as aforesaid, be enforceable by summary diligence".

(4) For subsection (3) of section thirty-eight there shall be substituted the following subsection:—

"(3) The Court of Session shall, subject to the provisions of this section, have power, exercisable by statutory instrument, to make provision by Act of Sederunt with respect to the evidence which must be furnished by a party seeking to enforce in Scotland an award under this Part of this Act, and the Statutory Instruments Act, 1946, shall apply to a statutory instrument containing an Act of Sederunt made under this subsection as if the Act of Sederunt has been made by a Minister of the Crown".

**42.** — (1) The following provisions of this section shall have effect for the purpose of the application of this Part of this Act to Northern Ireland.

(2) For the references to England there shall be substituted references to Northern Ireland.

[Sub-sections 42 (3) and (4) and section 43 repealed.]

# CONSTRUCTION, ENGLAND AND WALES
## The Scheme for Construction Contracts (England and Wales) Regulations 1998

### Statutory Instrument 1998 No. 649

The Secretary of State, in exercise of the powers conferred on him by sections 108(6), 114 and 146(1) and (2) of the Housing Grants, Construction and Regeneration Act 1996, and of all other powers enabling him in that behalf, having consulted such persons as he thinks fit, and draft Regulations having been approved by both Houses of Parliament, hereby makes the following Regulations:

### Citation, commencement, extent and interpretation

1.—(1) These Regulations may be cited as the Scheme for Construction Contracts (England and Wales) Regulations 1998 and shall come into force at the end of the period of 8 weeks beginning with the day on which they are made (the "commencement date").

(2) These Regulations shall extend only to England and Wales.

(3) In these Regulations, "the Act" means the Housing Grants, Construction and Regeneration Act 1996.

### The Scheme for Construction Contracts

2. Where a construction contract does not comply with the requirements of section 108(1) to (4) of the Act, the adjudication provisions in Part I of the Schedule to these Regulations shall apply.

3. Where—
   (a) the parties to a construction contract are unable to reach agreement for the purposes mentioned respectively in sections 109, 111 and 113 of the Act, or
   (b) a construction contract does not make provision as required by section 110 of the Act,

the relevant provisions in Part II of the Schedule to these Regulations shall apply.

**4.** The provisions in the Schedule to these Regulations shall be the Scheme for Construction Contracts for the purposes of section 114 of the Act.

Signed by authority of the Secretary of State

SCHEDULE          Regulations 2, 3 and 4

THE SCHEME FOR CONSTRUCTION CONTRACTS
PART I—ADJUDICATION

### Notice of Intention to seek Adjudication

**1.**—(1) Any party to a construction contract (the "referring party" may give written notice (the "notice of adjudication") of his intention to refer any dispute arising under the contract, to adjudication.

(2) The notice of adjudication shall be given to every other party to the contract.

(3) The notice of adjudication shall set out briefly—
   (a) the nature and a brief description of the dispute and of the parties involved,
   (b) details of where and when the dispute has arisen,
   (c) the nature of the redress which is sought, and
   (d) the names and addresses of the parties to the contract (including, where appropriate, the addresses which the parties have specified for the giving of notices).

**2.**—(1) Following the giving of a notice of adjudication and subject to any agreement between the parties to the dispute as to who shall act as adjudicator—
   (a) the referring party shall request the person (if any) specified in the contract to act as adjudicator, or
   (b) if no person is named in the contract or the person named has already indicated that he is unwilling or unable to act, and the contract provides for a specified nominating body to select a person, the referring party shall request the nominating body named in the contract to select a person to act as adjudicator, or
   (c) where neither paragraph (a) nor (b) above applies, or where the person referred to in (a) has already indicated that he is unwilling or unable to act and (b) does not apply, the referring party shall

request an adjudicator nominating body to select a person to act as adjudicator.

(2) A person requested to act as adjudicator in accordance with the provisions of paragraph (1) shall indicate whether or not he is willing to act within two days of receiving the request.

(3) In this paragraph, and in paragraphs 5 and 6 below, an "adjudicator nominating body" shall mean a body (not being a natural person and not being a party to the dispute) which holds itself out publicly as a body which will select an adjudicator when requested to do so by a referring party.

**3.** The request referred to in paragraphs 2, 5 and 6 shall be accompanied by a copy of the notice of adjudication.

**4.** Any person requested or selected to act as adjudicator in accordance with paragraphs 2, 5 or 6 shall be a natural person acting in his personal capacity. A person requested or selected to act as an adjudicator shall not be an employee of any of the parties to the dispute and shall declare any interest, financial or otherwise, in any matter relating to the dispute.

**5.**—(1) The nominating body referred to in paragraphs 2(1)(b) and 6(1)(b) or the adjudicator nominating body referred to in paragraphs 2(1)(c), 5(2)(b) and 6(1)(c) must communicate the selection of an adjudicator to the referring party within five days of receiving a request to do so.

(2) Where the nominating body or the adjudicator nominating body fails to comply with paragraph (1), the referring party may—
  (a) agree with the other party to the dispute to request a specified person to act as adjudicator, or
  (b) request any other adjudicator nominating body to select a person to act as adjudicator.

(3) The person requested to act as adjudicator in accordance with the provisions of paragraphs (1) or (2) shall indicate whether or not he is willing to act within two days of receiving the request.

**6.**—(1) Where an adjudicator who is named in the contract indicates to the parties that he is unable or unwilling to act, or where he fails to respond in accordance with paragraph 2(2), the referring party may—
  (a) request another person (if any) specified in the contract to act as adjudicator, or
  (b) request the nominating body (if any) referred to in the contract to select a person to act as adjudicator, or
  (c) request any other adjudicator nominating body to select a person to act as adjudicator.

(2) The person requested to act in accordance with the provisions of

paragraph (1) shall indicate whether or not he is willing to act within two days of receiving the request.

7.—(1) Where an adjudicator has been selected in accordance with paragraphs 2, 5 or 6, the referring party shall, not later than seven days from the date of the notice of adjudication, refer the dispute in writing (the "referral notice") to the adjudicator.

(2) A referral notice shall be accompanied by copies of, or relevant extracts from, the construction contract and such other documents as the referring party intends to rely upon.

(3) The referring party shall, at the same time as he sends to the adjudicator the documents referred to in paragraphs (1) and (2), send copies of those documents to every other party to the dispute.

8.—(1) The adjudicator may, with the consent of all the parties to those disputes, adjudicate at the same time on more than one dispute under the same contract.

(2) The adjudicator may, with the consent of all the parties to those disputes, adjudicate at the same time on related disputes under different contracts, whether or not one or more of those parties is a party to those disputes.

(3) All the parties in paragraphs (1) and (2) respectively may agree to extend the period within which the adjudicator may reach a decision in relation to all or any of these disputes.

(4) Where an adjudicator ceases to act because a dispute is to be adjudicated on by another person in terms of this paragraph, that adjudicator's fees and expenses shall be determined in accordance with paragraph 25.

9.—(1) An adjudicator may resign at any time on giving notice in writing to the parties to the dispute.

(2) An adjudicator must resign where the dispute is the same or substantially the same as one which has previously been referred to adjudication, and a decision has been taken in that adjudication.

(3) Where an adjudicator ceases to act under paragraph 9(1) —
  (a) the referring party may serve a fresh notice under paragraph 1 and shall request an adjudicator to act in accordance with paragraphs 2 to 7; and
  (b) if requested by the new adjudicator and insofar as it is reasonably practicable, the parties shall supply him with copies of all documents which they had made available to the previous adjudicator.

(4) Where an adjudicator resigns in the circumstances referred to in paragraph (2), or where a dispute varies significantly from the dispute referred to him in the referral notice and for that reason he is not competent

to decide it, the adjudicator shall be entitled to the payment of such reasonable amount as he may determine by way of fees and expenses reasonably incurred by him. The parties shall be jointly and severally liable for any sum which remains outstanding following the making of any determination on how the payment shall be apportioned.

10. Where any party to the dispute objects to the appointment of a particular person as adjudicator, that objection shall not invalidate the adjudicator's appointment nor any decision he may reach in accordance with paragraph 20.

11.—(1) The parties to a dispute may at any time agree to revoke the appointment of the adjudicator. The adjudicator shall be entitled to the payment of such reasonable amount as he may determine by way of fees and expenses incurred by him. The parties shall be jointly and severally liable for any sum which remains outstanding following the making of any determination on how the payment shall be apportioned.

(2) Where the revocation of the appointment of the adjudicator is due to the default or misconduct of the adjudicator, the parties shall not be liable to pay the adjudicator's fees and expenses.

**Powers of the adjudicator**

12. The adjudicator shall—
   (a) act impartially in carrying out his duties and shall do so in accordance with any relevant terms of the contract and shall reach his decision in accordance with the applicable law in relation to the contract; and
   (b) avoid incurring unnecessary expense.

13. The adjudicator may take the initiative in ascertaining the facts and the law necessary to determine the dispute, and shall decide on the procedure to be followed in the adjudication. In particular he may—
   (a) request any party to the contract to supply him with such documents as he may reasonably require including, if he so directs, any written statement from any party to the contract supporting or supplementing the referral notice and any other documents given under paragraph 7(2),
   (b) decide the language or languages to be used in the adjudication and whether a translation of any document is to be provided and if so by whom,
   (c) meet and question any of the parties to the contract and their representatives,
   (d) subject to obtaining any necessary consent from a third party or parties, make such site visits and inspections as he considers appropriate, whether accompanied by the parties or not,

   (e) subject to obtaining any necessary consent from a third party or parties, carry out any tests or experiments,

   (f) obtain and consider such representations and submissions as he requires, and, provided he has notified the parties of his intention, appoint experts, assessors or legal advisers,

   (g) give directions as to the timetable for the adjudication, any deadlines, or limits as to the length of written documents or oral representations to be complied with, and

   (h) issue other directions relating to the conduct of the adjudication.

**14.** The parties shall comply with any request or direction of the adjudicator in relation to the adjudication.

**15.** If, without showing sufficient cause, a party fails to comply with any request, direction or timetable of the adjudicator made in accordance with his powers, fails to produce any document or written statement requested by the adjudicator, or in any other way fails to comply with a requirement under these provisions relating to the adjudication, the adjudicator may —

   (a) continue the adjudication in the absence of that party or of the document or written statement requested,

   (b) draw such inferences from that failure to comply as circumstances may, in the adjudicator's opinion, be justified, and

   (c) make a decision on the basis of the information before him attaching such weight as he thinks fit to any evidence submitted to him outside any period he may have requested or directed.

**16.**—(1) Subject to any agreement between the parties to the contrary, and to the terms of paragraph (2) below, any party to the dispute may be assisted by, or represented by, such advisers or representatives (whether legally qualified or not) as he considers appropriate.

(2) Where the adjudicator is considering oral evidence or representations, a party to the dispute may not be represented by more than one person, unless the adjudicator gives directions to the contrary.

**17.** The adjudicator shall consider any relevant information submitted to him by any of the parties to the dispute and shall make available to them any information to be taken into account in reaching his decision.

**18.** The adjudicator and any party to the dispute shall not disclose to any other person any information or document provided to him in connection with the adjudication which the party supplying it has indicated is to be treated as confidential, except to the extent that it is necessary for the purposes of, or in connection with, the adjudication.

**19.**—(1) The adjudicator shall reach his decision not later than—

   (a) twenty eight days after the date of the referral notice mentioned in paragraph 7(1), or

(b) forty two days after the date of the referral notice if the referring party so consents, or

(c) such period exceeding twenty eight days after the referral notice as the parties to the dispute may, after the giving of that notice, agree.

(2) Where the adjudicator fails, for any reason, to reach his decision in accordance with paragraph (1)

(a) any of the parties to the dispute may serve a fresh notice under paragraph 1 and shall request an adjudicator to act in accordance with paragraphs 2 to 7; and

(b) if requested by the new adjudicator and insofar as it is reasonably practicable, the parties shall supply him with copies of all documents which they had made available to the previous adjudicator.

(3) As soon as possible after he has reached a decision, the adjudicator shall deliver a copy of that decision to each of the parties to the contract.

**Adjudicator's decision**

**20.** The adjudicator shall decide the matters in dispute. He may take into account any other matters which the parties to the dispute agree should be within the scope of the adjudication or which are matters under the contract which he considers are necessarily connected with the dispute. In particular, he may —

(a) open up, revise and review any decision taken or any certificate given by any person referred to in the contract unless the contract states that the decision or certificate is final and conclusive,

(b) decide that any of the parties to the dispute is liable to make a payment under the contract (whether in sterling or some other currency) and, subject to section 111(4) of the Act, when that payment is due and the final date for payment,

(c) having regard to any term of the contract relating to the payment of interest decide the circumstances in which, and the rates at which, and the periods for which simple or compound rates of interest shall be paid.

**21.** In the absence of any directions by the adjudicator relating to the time for perfomance of his decision, the parties shall be required to comply with any decision of the adjudicator immediately on delivery of the decision to the parties in accordance with this paragraph.

**22.** If requested by one of the parties to the dispute, the adjudicator shall provide reasons for his decision.

**Effects of the decision**

**23.**—(1) In his decision, the adjudicator may, if he thinks fit, order any of the parties to comply peremptorily with his decision or any part of it.

(2) The decision of the adjudicator shall be binding on the parties, and they shall comply with it until the dispute is finally determined by legal proceedings, by arbitration (if the contract provides for arbitration or the parties otherwise agree to arbitration) or by agreement between the parties.

**24.** Section 42 of the Arbitration Act 1996 shall apply to this Scheme subject to the following modifications —
   (a) in subsection (2) for the word "tribunal" wherever it appears there shall be substituted the word "adjudicator",
   (b) in subparagraph (b) of subsection (2) for the words "arbitral proceedings" there shall be substituted the word "adjudication",
   (c) subparagraph (c) of subsection (2) shall be deleted, and
   (d) subsection (3) shall be deleted.

**25.** The adjudicator shall be entitled to the payment of such reasonable amount as he may determine by way of fees and expenses reasonably incurred by him. The parties shall be jointly and severally liable for any sum which remains outstanding following the making of any determination on how the payment shall be apportioned.

**26.** The adjudicator shall not be liable for anything done or omitted in the discharge or purported discharge of his functions as adjudicator unless the act or omission is in bad faith, and any employee or agent of the adjudicator shall be similarly protected from liability.

PART II
PAYMENT

## Entitlement to and amount of stage payments

1. Where the parties to a relevant construction contract fail to agree —
   (a) the amount of any instalment or stage or periodic payment for any work under the contract, or
   (b) the intervals at which, or circumstances in which, such payments become due under that contract, or
   (c) both of the matters mentioned in sub-paragraphs (a) and (b) above,
the relevant provisions of paragraphs 2 to 4 below shall apply.

2. — (1) The amount of any payment by way of instalments or stage or periodic payments in respect of a relevant period shall be the difference between the amount determined in accordance with sub-paragraph (2) and the amount determined in accordance with sub-paragraph (3).

(2) The aggregate of the following amounts —
   (a) an amount equal to the value of any work performed in accordance with the relevant construction contract during the period

246

from the commencement of the contract to the end of the relevant period (excluding any amount calculated in accordance with subparagraph (b)),

(b) where the contract provides for payment for materials, an amount equal to the value of any materials manufactured on site or brought onto site for the purposes of the works during the period from the commencement of the contract to the end of the relevant period, and

(c) any other amount or sum which the contract specifies shall be payable during or in respect of the period from the commencement of the contract to the end of the relevant period.

(3) The aggregate of any sums which have been paid or are due for payment by way of instalments, stage or periodic payments during the period from the commencement of the contract to the end of the relevant period.

(4) An amount calculated in accordance with this paragraph shall not exceed the difference between—

(a) the contract price, and

(b) the aggregate of the instalments or stage or periodic payments which have become due.

## Dates for payment

3. Where the parties to a construction contract fail to provide an adequate mechanism for determining either what payments become due under the contract, or when they become due for payment, or both, the relevant provisions of paragraphs 4 to 7 shall apply.

4. Any payment of a kind mentioned in paragraph 2 above shall become due on whichever of the following dates occurs later—

(a) the expiry of 7 days following the relevant period mentioned in paragraph 2(1) above, or

(b) the making of a claim by the payee.

5. The final payment payable under a relevant construction contract, namely the payment of an amount equal to the difference (if any) between—

(a) the contract price, and

(b) the aggregate of any instalment or stage or periodic payments which have become due under the contract,

shall become due on the expiry of—

(a) 30 days following completion of the work, or

(b) the making of a claim by the payee,

whichever is the later.

6. Payment of the contract price under a construction contract (not being a relevant construction contract) shall become due on

(a) the expiry of 30 days following the completion of the work, or
(b) the making of a claim by the payee,
whichever is the later.

**7.** Any other payment under a construction contract shall become due
(a) on the expiry of 7 days following the completion of the work to which the payment relates, or
(b) the making of a claim by the payee,
whichever is the later.

## Final date for payment

**8.** — (1) Where the parties to a construction contract fail to provide a final date for payment in relation to any sum which becomes due under a construction contract, the provisions of this paragraph shall apply.

(2) The final date for the making of any payment of a kind mentioned in paragraphs 2, 5, 6 or 7, shall be 17 days from the date that payment becomes due.

## Notice specifying amount of payment

**9.** A party to a construction contract shall, not later than 5 days after the date on which any payment —
(a) becomes due from him, or
(b) would have become due, if —
    (i) the other party had carried out his obligations under the contract and
    (ii) no set-off or abatement was permitted by reference to any sum claimed to be due under one or more other contracts,
give notice to the other party to the contract specifying the amount (if any) of the payment he has made or proposes to make, specifying to what the payment relates and the basis on which that amount is calculated.

## Notice of intention to withhold payment

**10.** Any notice of intention to withhold payment mentioned in section 111 of the Act shall be given not later than the prescribed period, which is to say not later than 7 days before the final date for payment determined either in accordance with the construction contract, or where no such provision is made in the contract, in accordance with paragraph 8 above.

## Prohibition of conditional payment provisions

**11.** Where a provision making payment under a construction contract conditional on the payer receiving payment from a third person is ineffective as mentioned in section 113 of the Act, and the parties have not agreed other terms for payment, the relevant provisions of —

(a) paragraphs 2, 4, 5, 7, 8, 9 and 10 shall apply in the case of a relevant construction contract, and

(b) paragraphs 6, 7, 8, 9 and 10 shall apply in the case of any other construction contract.

## Interpretation

**12.** In this Part of the Scheme for Construction Contracts—

"claim by the payee" means a written notice given by the party carrying out work under a construction contract to the other party specifying the amount of any payment or payments which he considers to be due and the basis on which it is, or they are calculated;

"contract price" means the entire sum payable under the construction contract in respect of the work;

"relevant construction contract" means any construction contract other than one—

(a) which specifies that the duration of the work is to be less than 45 days, or

(b) in respect of which the parties agree that the duration of the work is estimated to be less than 45 days;

"relevant period" means a period which is specified in, or is calculated by reference to the construction contract or where no such period is so specified or is so calculable, a period of 28 days;

"value of work" means an amount determined in accordance with the construction contract under which the work is performed or where the contract contains no such provision, the cost of any work performed in accordance with that contract together with an amount equal to any overhead or profit included in the contract price;

"work" means any of the work or services mentioned in section 104 of the Act.

# EXPLANATORY NOTE

*(This note is not part of the Order)*

Part II of the Housing Grants, Construction and Regeneration Act 1996 makes provision in relation to construction contracts. Section 114 empowers the Secretary of State to make the Scheme for Construction Contracts. Where a construction contract does not comply with the requirements of sections 108 to 111 (adjudication of disputes and payment provisions), and section 113 (prohibition of conditional payment provisions), the relevant provisions of the Scheme for Construction Contracts have effect.

The Scheme which is contained in the Schedule to these Regulations is in two parts. Part I provides for the selection and appointment of an adjudicator, gives powers to the adjudicator to gather and consider information, and makes provisions in respect of his decisions. Part II makes provision with respect to payments under a construction contract where either the contract fails to make provision or the parties fail to agree—

(a) the method for calculating the amount of any instalment, stage or periodic payment,

(b) the due date and the final date for payments to be made, and

(c) prescribes the period within which a notice of intention to withhold payment must be given.

# APPENDIX E
# THE UNFAIR TERMS IN CONSUMER CONTRACTS REGULATIONS 1999

## Statutory Instrument 1999 No. 2083

### Citation and commencement

**1.** These Regulations may be cited as the Unfair Terms in Consumer Contracts Regulations 1999 and shall come into force on 1st October 1999.

### Revocation

**2.** The Unfair Terms in Consumer Contracts Regulations 1994 are hereby revoked.

### Interpretation

**3.**–(1) In these Regulations—
"the Community" means the European Community;
"consumer" means a natural person who, in contracts covered by these Regulations, is acting for purposes which are outside his trade, business or profession;
"court" in relation to England and Wales and Northern Ireland means a county court or the High Court, and in relation to Scotland, the Court of Session;
"Director" means the Director General of Fair Trading;
"EEA Agreement" means the Agreement on the European Economic Area signed at Oporto on 2nd May 1992 as adjusted by the protocol signed at Brussels on 17th March 1993;
"Member State" means a State which is a contracting party to the EEA Agreement;
"notified" means notified in writing;
"qualifying body" means a person specified in Schedule 1;
"seller or supplier" means any natural or legal person who, in contracts covered by these Regulations, is acting for purposes relating to his trade, business or profession, whether publicly owned or privately owned;
"unfair terms" means the contractual terms referred to in regulation 5.

(2) In the application of these Regulations to Scotland for references to

an "injunction" or an "interim injunction" there shall be substituted references to an "interdict" or "interim interdict" respectively.

## Terms to which these Regulations apply

**4.** — (1) These Regulations apply in relation to unfair terms in contracts concluded between a seller or a supplier and a consumer.

(2) These Regulations do not apply to contractual terms which reflect —
- (a) mandatory statutory or regulatory provisions (including such provisions under the law of any Member State or in Community legislation having effect in the United Kingdom without further enactment);
- (b) the provisions or principles of international conventions to which the Member States or the Community are party.

## Unfair terms

**5.** — (1) A contractual term which has not been individually negotiated shall be regarded as unfair if, contrary to the requirement of good faith, it causes a significant imbalance in the parties' rights and obligations arising under the contract, to the detriment of the consumer.

(2) A term shall always be regarded as not having been individually negotiated where it has been drafted in advance and the consumer has therefore not been able to influence the substance of the term.

(3) Notwithstanding that a specific term or certain aspects of it in a contract has been individually negotiated, these Regulations shall apply to the rest of a contract if an overall assessment of it indicates that it is a pre-formulated standard contract.

(4) It shall be for any seller or supplier who claims that a term was individually negotiated to show that it was.

(5) Schedule 2 to these Regulations contains an indicative and non-exhaustive list of the terms which may be regarded as unfair.

## Assessment of unfair terms

**6.** — (1) Without prejudice to regulation 12, the unfairness of a contractual term shall be assessed, taking into account the nature of the goods or services for which the contract was concluded and by referring, at the time of conclusion of the contract, to all the circumstances attending the conclusion of the contract and to all the other terms of the contract or of another contract on which it is dependent.

(2) In so far as it is in plain intelligible language, the assessment of fairness of a term shall not relate —
- (a) to the definition of the main subject matter of the contract, or
- (b) to the adequacy of the price or remuneration, as against the goods or services supplied in exchange.

## Written contracts

7.—(1) A seller or supplier shall ensure that any written term of a contract is expressed in plain, intelligible language

(2) If there is doubt about the meaning of a written term, the interpretation which is most favourable to the consumer shall prevail but this rule shall not apply in proceedings brought under regulation 12.

## Effect of unfair term

8.—(1) An unfair term in a contract concluded with a consumer by a seller or supplier shall not be binding on the consumer.

(2) The contract shall continue to bind the parties if it is capable of continuing in existence without the unfair term.

## Choice of law clauses

9. These Regulations shall apply notwithstanding any contract term which applies or purports to apply the law of a non-Member State, if the contract has a close connection with the territory of the Member States.

## Complaints – consideration by Director

10.—(1) It shall be the duty of the Director to consider any complaint made to him that any contract term drawn up for general use is unfair, unless—
  (a) the complaint appears to the Director to be frivolous or vexatious; or
  (b) a qualifying body has notified the Director that it agrees to consider the complaint.

(2) The Director shall give reasons for his decision to apply or not to apply, as the case may be, for an injunction under regulation 12 in relation to any complaint which these Regulations require him to consider.

(3) In deciding whether or not to apply for an injunction in respect of a term which the Director considers to be unfair, he may, if he considers it appropriate to do so, have regard to any undertakings given to him by or on behalf of any person as to the continued use of such a term in contracts concluded with consumers.

## Complaints – consideration by qualifying bodies

11.—(1) If a qualifying body specified in Part One of Schedule 1 notifies the Director that it agrees to consider a complaint that any contract term drawn up for general use is unfair, it shall be under a duty to consider that complaint.

(2) Regulation 10(2) and (3) shall apply to a qualifying body which is under a duty to consider a complaint as they apply to the Director.

### Injunctions to prevent continued use of unfair terms

**12.**—(1) The Director or, subject to paragraph (2), any qualifying body may apply for an injunction (including an interim injunction) against any person appearing to the Director or that body to be using, or recommending use of, an unfair term drawn up for general use in contracts concluded with consumers.

(2) A qualifying body may apply for an injunction only where—
  (a) it has notified the Director of its intention to apply at least fourteen days before the date on which the application is made, beginning with the date on which the notification was given; or
  (b) the Director consents to the application being made within a shorter period.

(3) The court on an application under this regulation may grant an injunction on such terms as it thinks fit.

(4) An injunction may relate not only to use of a particular contract term drawn up for general use but to any similar term, or a term having like effect, used or recommended for use by any person.

### Powers of the Director and qualifying bodies to obtain documents and information

**13.**—(1) The Director may exercise the power conferred by this regulation for the purpose of—
  (a) facilitating his consideration of a complaint that a contract term drawn up for general use is unfair; or
  (b) ascertaining whether a person has complied with an undertaking or court order as to the continued use, or recommendation for use, of a term in contracts concluded with consumers.

(2) A qualifying body specified in Part One of Schedule 1 may exercise the power conferred by this regulation for the purpose of—
  (a) facilitating its consideration of a complaint that a contract term drawn up for general use is unfair; or
  (b) ascertaining whether a person has complied with—
    (i) an undertaking given to it or to the court following an application by that body, or
    (ii) a court order made on an application by that body,
  as to the continued use, or recommendation for use, of a term in contracts concluded with consumers.

(3) The Director may require any person to supply to him, and a qualifying body specified in Part One of Schedule 1 may require any person to supply to it—
  (a) a copy of any document which that person has used or recommended for use, at the time the notice referred to in paragraph (4)

below is given, as a pre-formulated standard contract in dealings with consumers;

  (b) information about the use, or recommendation for use, by that person of that document or any other such document in dealings with consumers.

(4) The power conferred by this regulation is to be exercised by a notice in writing which may—

  (a) specify the way in which and the time within which it is to be complied with; and

  (b) be varied or revoked by a subsequent notice.

(5) Nothing in this regulation compels a person to supply any document or information which he would be entitled to refuse to produce or give in civil proceedings before the court.

(6) If a person makes default in complying with a notice under this regulation, the court may, on the application of the Director or of the qualifying body, make such order as the court thinks fit for requiring the default to be made good, and any such order may provide that all the costs or expenses of and incidental to the application shall be borne by the person in default or by any officers of a company or other association who are responsible for its default.

## Notification of undertakings and orders to Director

14. A qualifying body shall notify the Director—

  (a) of any undertaking given to it by or on behalf of any person as to the continued use of a term which that body considers to be unfair in contracts concluded with consumers;

  (b) of the outcome of any application made by it under regulation 12, and of the terms of any undertaking given to, or order made by, the court;

  (c) of the outcome of any application made by it to enforce a previous order of the court.

## Publication, information and advice

15.—(1) The Director shall arrange for the publication in such form and manner as he considers appropriate, of—

  (a) details of any undertaking or order notified to him under regulation 14;

  (b) details of any undertaking given to him by or on behalf of any person as to the continued use of a term which the Director considers to be unfair in contracts concluded with consumers;

  (c) details of any application made by him under regulation 12, and of the terms of any undertaking given to, or order made by, the court;

  (d) details of any application made by the Director to enforce a previous order of the court.

(2) The Director shall inform any person on request whether a particular term to which these Regulations apply has been—

    (a) the subject of an undertaking given to the Director or notified to him by a qualifying body; or

    (b) the subject of an order of the court made upon application by him or notified to him by a qualifying body;

and shall give that person details of the undertaking or a copy of the order, as the case may be, together with a copy of any amendments which the person giving the undertaking has agreed to make to the term in question.

(3) The Director may arrange for the dissemination in such form and manner as he considers appropriate of such information and advice concerning the operation of these Regulations as may appear to him to be expedient to give to the public and to all persons likely to be affected by these Regulations.

# SCHEDULE 1

Regulation 3

## QUALIFYING BODIES

### PART ONE

1. The Data Protection Registrar.

2. The Director General of Electricity Supply.

3. The Director General of Gas Supply.

4. The Director General of Electricity Supply for Northern Ireland.

5. The Director General of Gas for Northern Ireland.

6. The Director General of Telecommunications.

7. The Director General of Water Services.

8. The Rail Regulator.

9. Every weights and measures authority in Great Britain.

10. The Department of Economic Development in Northern Ireland.

### PART TWO

11. Consumers' Association.

SCHEDULE 2         Regulation 5(5)

# INDICATIVE AND NON-EXHAUSTIVE LIST OF TERMS WHICH MAY BE REGARDED AS UNFAIR

**1.** Terms which have the object or effect of—

  (a)  excluding or limiting the legal liability of a seller or supplier in the event of the death of a consumer or personal injury to the latter resulting from an act or omission of that seller or supplier;

  (b)  inappropriately excluding or limiting the legal rights of the consumer vis-à-vis the seller or supplier or another party in the event of total or partial non-performance or inadequate performance by the seller or supplier of any of the contractual obligations, including the option of offsetting a debt owed to the seller or supplier against any claim which the consumer may have against him;

  (c)  making an agreement binding on the consumer whereas provision of services by the seller or supplier is subject to a condition whose realisation depends on his own will alone;

  (d)  permitting the seller or supplier to retain sums paid by the consumer where the latter decides not to conclude or perform the contract, without providing for the consumer to receive compensation of an equivalent amount from the seller or supplier where the latter is the party cancelling the contract;

  (e)  requiring any consumer who fails to fulfil his obligation to pay a disproportionately high sum in compensation;

  (f)  authorising the seller or supplier to dissolve the contract on a discretionary basis where the same facility is not granted to the consumer, or permitting the seller or supplier to retain the sums paid for services not yet supplied by him where it is the seller or supplier himself who dissolves the contract;

  (g)  enabling the seller or supplier to terminate a contract of indeterminate duration without reasonable notice except where there are serious grounds for doing so;

  (h)  automatically extending a contract of fixed duration where the consumer does not indicate otherwise, when the deadline fixed for the consumer to express his desire not to extend the contract is unreasonably early;

  (i)  irrevocably binding the consumer to terms with which he had no real opportunity of becoming acquainted before the conclusion of the contract;

  (j)  enabling the seller or supplier to alter the terms of the contract unilaterally without a valid reason which is specified in the contract;

  (k)  enabling the seller or supplier to alter unilaterally without a valid reason any characteristics of the product or service to be provided;

(l)   providing for the price of goods to be determined at the time of delivery or allowing a seller of goods or supplier of services to increase their price without in both cases giving the consumer the corresponding right to cancel the contract if the final price is too high in relation to the price agreed when the contract was concluded:

(m)   giving the seller or supplier the right to determine whether the goods or services supplied are in conformity with the contract, or giving him the exclusive right to interpret any term of the contract;

(n)   limiting the seller's or supplier's obligation to respect commitments undertaken by his agents or making his commitments subject to compliance with a particular formality;

(o)   obliging the consumer to fulfil all his obligations where the seller or supplier does not perform his;

(p)   giving the seller or supplier the possibility of transferring his rights and obligations under the contract, where this may serve to reduce the guarantees for the consumer, without the latter's agreement;

(q)   excluding or hindering the consumer's right to take legal action or exercise any other legal remedy, particularly by requiring the consumer to take disputes exclusively to arbitration not covered by legal provisions, unduly restricting the evidence available to him or imposing on him a burden of proof which, according to the applicable law, should lie with another party to the contract.

2. Scope of subparagraphs 1(g), (j) and (l)

(a) Paragraph 1(g) is without hindrance to terms by which a supplier of financial services reserves the right to terminate unilaterally a contract of indeterminate duration without notice where there is a valid reason, provided that the supplier is required to inform the other contracting party or parties thereof immediately.

(b) Paragraph 1(j) is without hindrance to terms under which a supplier of financial services reserves the right to alter the rate of interest payable by the consumer or due to the latter, or the amount of other charges for financial services without notice where there is a valid reason, provided that the supplier is required to inform the other contracting party or parties thereof at the earliest opportunity and that the latter are free to dissolve the contract immediately.

Paragraph 1(j) is also without hindrance to terms under which a seller or supplier reserves the right to alter unilaterally the conditions of a contract of indeterminate duration, provided that he is required to inform the consumer with reasonable notice and that the consumer is free to dissolve the contract.

(c) Subparagraphs 1(g), (j) and (l) do not apply to:
  − transactions in transferable securities, financial instruments and other products or services where the price is linked to

fluctuations in a stock exchange quotation or index or a financial market rate that the seller or supplier does not control;

— contracts for the purchase or sale of foreign currency, traveller's cheques or international money orders denominated in foreign currency;

(d) Paragraph 1(l) is without hindrance to price indexation clauses, where lawful, provided that the method by which prices vary is explicitly described.

## EXPLANATORY NOTE

*(This note is not part of the Regulations)*

These Regulations revoke and replace the Unfair Terms in Consumer Contracts Regulations 1994 (S.I. 1994/3159) which came into force on 1st July 1995.

Those Regulations implemented Council Directive 93/13/EEC on unfair terms in consumer contracts (O.J. No. L95, 21.4.93, p. 29). Regulations 3 to 9 of these Regulations re-enact regulations 2 to 7 of the 1994 Regulations with modifications to reflect more closely the wording of the Directive.

The Regulations apply, with certain exceptions, to unfair terms in contracts concluded between a consumer and a seller or supplier (regulation 4). The Regulations provide that an unfair term is one which has not been individually negotiated and which, contrary to the requirement of good faith, causes a significant imbalance in the parties' rights and obligations under the contract to the detriment of the consumer (regulation 5). Schedule 2 contains an indicative list of terms which may be regarded as unfair.

The assessment of unfairness will take into account all the circumstances attending the conclusion of the contract. However, the assessment is not to relate to the definition of the main subject matter of the contract or the adequacy of the price or remuneration as against the goods or services supplied in exchange as long as the terms concerned are in plain, intelligible language (regulation 6). Unfair contract terms are not binding on the consumer (regulation 8).

The Regulations maintain the obligation on the Director General of Fair Trading (contained in the 1994 Regulations) to consider any complaint made to him about the fairness of any contract term drawn up for general use. He may, if he considers it appropriate to do so, seek an injunction to prevent the continued use of that term or of a term having like effect (regulations 10 and 12).

The Regulations provide for the first time that a qualifying body named in Schedule 1 (statutory regulators, trading standards departments and Consumers' Association) may also apply for an injunction to prevent the continued use of an unfair contract term provided it has notified the Director General of its intention at least 14 days before the application is made (unless the Director General consents to a shorter period) (regulation 12). A qualifying body named in Part One of Schedule 1 (public bodies) shall be under a duty to consider a complaint if it has told the Director General that it will do so (regulation 11).

The Regulations provide a new power for the Director General and the public qualifying bodies to require traders to produce copies of their standard contracts, and give information about their use, in order to facilitate investigation of complaints and ensure compliance with undertakings or court orders (regulation 13).

Qualifying bodies must notify the Director General of undertakings given to them about the continued use of an unfair term and of the outcome of any court proceedings (regulation 14). The Director General is given the power to arrange for the publication of this information in such form and manner as he considers appropriate and to offer information and advice about the operation of these Regulations (regulation 15). In addition the Director General will supply enquirers about particular standard terms with details of any relevant undertakings and court orders.

A Regulatory Impact Assessment of the costs and benefits which will result from these Regulations has been prepared by the Department of Trade and Industry and is available from Consumer Affairs Directorate, Department of Trade and Industry, Room 407, 1 Victoria Street, London SW1H 0ET (Telephone 020 7215 0341). Copies have been placed in the libraries of both Houses of Parliament.

# APPENDIX F
# THE UNFAIR ARBITRATION AGREEMENTS (SPECIFIED AMOUNT) ORDER 1999

## Statutory Instrument 1999 No. 2167

The Secretary of State, in exercise of the power conferred on him by section 91(1) and 3(a) and (b) of the Arbitration Act 1996, with the concurrence (as respects England and Wales) of the Lord Chancellor, hereby makes the following Order:

**1.** This Order may be cited as the Unfair Arbitration Agreements (Specified Amount) Order 1999, and shall come into force on 1st January 2000.

**2.** The Unfair Arbitration Agreements (Specified Amount) Order 1996 is hereby revoked.

**3.** The amount of £5,000 is hereby specified for the purposes of section 91 of the Arbitration Act 1996 (arbitration agreement unfair where modest amount sought).

## EXPLANATORY NOTE
### (This note is not part of the Order)

This Order specifies the amount of £5,000 for the purposes of section 91 of the Arbitration Act 1996 for England and Wales and Scotland. It replaces The Unfair Arbitration Agreements (Specified Amount) Order 1996 which specified the amount of £3,000.

Section 91(1) of the 1996 Act provides that a term which constitutes an arbitration agreement is unfair for the purposes of the Unfair Terms in Consumer Contracts Regulations 1994 (S.I. 1994/3159) so far as it relates to a claim for a pecuniary remedy which does not exceed the amount specified by order for the purposes of that section. The 1994 Regulations implement Council Directive 93/13/EEC on unfair terms in consumer contracts (O.J. No. L95, 21.4.93, p. 29). The Unfair Terms in Consumer

Contracts Regulations 1999 (S.I. 1999/2083) revoke and replace the 1994 Regulations with effect from 1st October 1999.

Section 89(1) of the 1996 Act defines "arbitration agreement" as an agreement to submit to arbitration present or future disputes or differences (whether or not contractual).

A Regulatory Impact Assessment of the costs and benefits which will result from this Order has been prepared by the Department of Trade and Industry and is available from Consumer Affairs Directorate, Department of Trade and Industry, Room 407, 1 Victoria Street, London SW1H 0ET (telephone 020 7215 0342). Copies have been placed in the libraries of both Houses of Parliament.

# APPENDIX G
# INTEREST TABLES

## Table 1: *Base rates*

| Date from | Base Rate % | Date from | Base Rate |
|---|---|---|---|
| 05.10.89 | 15.00 | 06.05.97 | 06.25 |
| 08.10.90 | 14.00 | 06.06.97 | 06.50 |
| 13.02.91 | 13.50 | 11.07.97 | 06.75 |
| 27.02.91 | 13.00 | 08.08.97 | 07.00 |
| 22.03.91 | 12.50 | 06.11.97 | 07.25 |
| 12.04.91 | 12.00 | 04.06.98 | 07.50 |
| 24.05.91 | 11.50 | 08.10.98 | 07.25 |
| 13.07.91 | 11.00 | 05.11.98 | 06.75 |
| 04.09.91 | 10.50 | 10.12.98 | 06.25 |
| 05.05.92 | 10.00 | 07.01.99 | 06.00 |
| 17.09.92 | 12.00 | 04.02.99 | 05.50 |
| 18.09.92 | 10.00 | 08.04.99 | 05.25 |
| 23.09.92 | 09.00 | 10.06.99 | 05.00 |
| 17.10.92 | 08.00 | 09.09.99 | 05.25 |
| 13.11.92 | 07.00 | 04.11.99 | 05.50 |
| 26.01.93 | 06.00 | 13.01.00 | 05.75 |
| 23.11.93 | 05.50 | 10.02.00 | 06.00 |
| 08.02.94 | 05.25 | | |
| 13.09.94 | 05.75 | | |
| 08.12.94 | 06.25 | | |
| 02.02.95 | 06.75 | | |
| 13.12.95 | 06.50 | | |
| 19.01.96 | 06.25 | | |
| 08.03.96 | 06.00 | | |
| 07.06.96 | 05.75 | | |
| 30.10.96 | 06.00 | | |

## Table 2: *Simple interest at 2% above base rate*

| Date from | Base rate + 2% | Period (days) | Interest (% of principal) |
|---|---|---|---|
| 05.10.89 | 17.00 | 368 | 17.1397260 |
| 08.10.90 | 16.00 | 128 | 5.6109589 |
| 13.02.91 | 15.50 | 14 | 0.5945205 |
| 27.02.91 | 15.00 | 23 | 0.9452055 |
| 22.03.91 | 14.50 | 21 | 0.8342466 |
| 12.04.91 | 14.00 | 42 | 1.6109589 |
| 24.05.91 | 13.50 | 50 | 1.8493151 |
| 13.07.91 | 13.00 | 53 | 1.8876712 |
| 04.09.91 | 12.50 | 244 | 8.3444681 |
| 05.05.92 | 12.00 | 135 | 4.4262295 |
| 17.09.92 | 14.00 | 1 | 0.0382514 |
| 18.09.92 | 12.00 | 5 | 0.1639344 |
| 23.09.92 | 11.00 | 24 | 0.7213115 |
| 17.10.92 | 10.00 | 27 | 0.7377049 |
| 13.11.92 | 09.00 | 74 | 1.8213564 |
| 26.01.93 | 08.00 | 301 | 6.5972603 |
| 23.11.93 | 07.50 | 77 | 1.5821918 |
| 08.02.94 | 07.25 | 217 | 4.3102740 |
| 13.09.94 | 07.75 | 86 | 1.8260274 |
| 08.12.94 | 08.25 | 56 | 1.2657534 |
| 02.02.95 | 08.75 | 314 | 7.5273973 |
| 13.12.95 | 08.50 | 37 | 0.8604985 |
| 19.01.96 | 08.25 | 49 | 1.1045082 |
| 08.03.96 | 08.00 | 91 | 1.9890710 |
| 07.06.96 | 07.75 | 145 | 3.0703552 |
| 30.10.96 | 08.00 | 188 | 4.1167752 |
| 06.05.97 | 08.25 | 31 | 0.7006849 |
| 06.06.97 | 08.50 | 35 | 0.8150685 |
| 11.07.97 | 08.75 | 28 | 0.6712329 |
| 08.08.97 | 09.00 | 90 | 2.2191781 |
| 06.11.97 | 09.25 | 210 | 5.3219178 |
| 04.06.98 | 09.50 | 126 | 3.2794521 |
| 08.10.98 | 09.25 | 28 | 0.7095890 |
| 05.11.98 | 08.75 | 35 | 0.8390411 |
| 10.12.98 | 08.25 | 28 | 0.6328767 |
| 07.01.99 | 08.00 | 28 | 0.6136986 |
| 04.02.99 | 07.50 | 63 | 1.2945205 |
| 08.04.99 | 07.25 | 63 | 1.2513699 |
| 10.06.99 | 07.00 | 91 | 1.7452055 |
| 09.09.99 | 07.25 | 56 | 1.1123288 |
| 04.11.99 | 07.50 | 70 | 1.4376825 |
| 13.01.00 | 07.75 | 28 | 0.5928962 |
| 10.02.00 | 08.00 | | |

## Table 3: Interest at 2% above base rate compounded monthly

| Month/year | Mean base rate + 2% | Compound interest factor | Cumulative factor |
|---|---|---|---|
| Jan 90 | 17.0000000 | 1.0144384 | 1.0144384 |
| Feb 90 | 17.0000000 | 1.0130411 | 1.0276677 |
| Mar 90 | 17.0000000 | 1.0144384 | 1.0425056 |
| Apr 90 | 17.0000000 | 1.0139726 | 1.0570721 |
| May 90 | 17.0000000 | 1.0144384 | 1.0723345 |
| Jun 90 | 17.0000000 | 1.0139726 | 1.0873178 |
| Jul 90 | 17.0000000 | 1.0144384 | 1.1030169 |
| Aug 90 | 17.0000000 | 1.0144384 | 1.1189426 |
| Sep 90 | 17.0000000 | 1.0139726 | 1.1345772 |
| Oct 90 | 16.2258064 | 1.0137808 | 1.1502126 |
| Nov 90 | 16.0000000 | 1.0131507 | 1.1653386 |
| Dec 90 | 16.0000000 | 1.0135890 | 1.1811745 |
| Jan 91 | 16.0000000 | 1.0135890 | 1.1972255 |
| Feb 91 | 15.6785714 | 1.0120274 | 1.2116250 |
| Mar 91 | 14.8387097 | 1.0126027 | 1.2268948 |
| Apr 91 | 14.1833333 | 1.0116575 | 1.2411974 |
| May 91 | 13.8709677 | 1.0117808 | 1.2558197 |
| Jun 91 | 13.5000000 | 1.0110959 | 1.2697541 |
| Jul 91 | 13.1935484 | 1.0112055 | 1.2839823 |
| Aug 91 | 13.0000000 | 1.0110411 | 1.2981589 |
| Sep 91 | 12.5500000 | 1.0103151 | 1.3115495 |
| Oct 91 | 12.5000000 | 1.0106164 | 1.3254735 |
| Nov 91 | 12.5000000 | 1.0102740 | 1.3390914 |
| Dec 91 | 12.5000000 | 1.0106164 | 1.3533078 |
| Jan 92 | 12.5000000 | 1.0105874 | 1.3676358 |
| Feb 92 | 12.5000000 | 1.0099044 | 1.3811814 |
| Mar 92 | 12.5000000 | 1.0105874 | 1.3958045 |
| Apr 92 | 12.5000000 | 1.0102459 | 1.4101058 |
| May 92 | 12.0645161 | 1.0102186 | 1.4245151 |
| Jun 92 | 12.0000000 | 1.0098361 | 1.4385267 |
| Jul 92 | 12.0000000 | 1.0101639 | 1.4531478 |
| Aug 92 | 12.0000000 | 1.0101639 | 1.4679175 |
| Sep 92 | 11.8000000 | 1.0096721 | 1.4821154 |
| Oct 92 | 10.5161290 | 1.0089071 | 1.4953168 |
| Nov 92 | 09.4000000 | 1.0077049 | 1.5068381 |
| Dec 92 | 09.0000000 | 1.0076230 | 1.5183246 |
| Jan 93 | 08.8064516 | 1.0074795 | 1.5296808 |
| Feb 93 | 08.0000000 | 1.0061370 | 1.5390685 |
| Mar 93 | 08.0000000 | 1.0067945 | 1.5495257 |
| Apr 93 | 08.0000000 | 1.0065753 | 1.5597144 |

| Month/year | Mean basic rate + 2% | Compound interest factor | Cumulative factor |
|---|---|---|---|
| May 93 | 08.0000000 | 1.0067945 | 1.5703119 |
| Jun 93 | 08.0000000 | 1.0065753 | 1.5806372 |
| Jul 93 | 08.0000000 | 1.0067945 | 1.5913769 |
| Aug 93 | 08.0000000 | 1.0067945 | 1.6021895 |
| Sep 93 | 08.0000000 | 1.0065753 | 1.6127245 |
| Oct 93 | 08.0000000 | 1.0067945 | 1.6236822 |
| Nov 93 | 07.8666667 | 1.0064658 | 1.6341805 |
| Dec 93 | 07.5000000 | 1.0063699 | 1.6445900 |
| Jan 94 | 07.5000000 | 1.0063699 | 1.6550658 |
| Feb 94 | 07.3125000 | 1.0056096 | 1.6643501 |
| Mar 94 | 07.2500000 | 1.0061575 | 1.6745983 |
| Apr 94 | 07.2500000 | 1.0059589 | 1.6845771 |
| May 94 | 07.2500000 | 1.0061575 | 1.6949500 |
| Jun 94 | 07.2500000 | 1.0059589 | 1.7050500 |
| Jul 94 | 07.2500000 | 1.0061575 | 1.7155489 |
| Aug 94 | 07.2500000 | 1.0061575 | 1.7261125 |
| Sep 94 | 07.5500000 | 1.0062055 | 1.7368238 |
| Oct 94 | 07.7500000 | 1.0065822 | 1.7482559 |
| Nov 94 | 07.7500000 | 1.0063699 | 1.7593921 |
| Dec 94 | 08.1370968 | 1.0069110 | 1.7715512 |
| Jan 95 | 08.2500000 | 1.0070068 | 1.7839641 |
| Feb 95 | 08.7321429 | 1.0066986 | 1.7959143 |
| Mar 95 | 08.7500000 | 1.0074315 | 1.8092606 |
| Apr 95 | 08.7500000 | 1.0071918 | 1.8222724 |
| May 95 | 08.7500000 | 1.0074315 | 1.8358146 |
| Jun 95 | 08.7500000 | 1.0071918 | 1.8490174 |
| Jul 95 | 08.7500000 | 1.0074315 | 1.8627584 |
| Aug 95 | 08.7500000 | 1.0074315 | 1.8766015 |
| Sep 95 | 08.7500000 | 1.0071918 | 1.8900976 |
| Oct 95 | 08.7500000 | 1.0074315 | 1.9041439 |
| Nov 95 | 08.7500000 | 1.0071918 | 1.9178381 |
| Dec 95 | 08.5967742 | 1.0073014 | 1.9318409 |
| Jan 96 | 08.3951613 | 1.0071107 | 1.9455776 |
| Feb 96 | 08.2500000 | 1.0065369 | 1.9582956 |
| Mar 96 | 08.0564516 | 1.0068238 | 1.9716586 |
| Apr 96 | 08.0000000 | 1.0065574 | 1.9845875 |
| May 96 | 08.0000000 | 1.0067760 | 1.9980349 |
| Jun 96 | 07.8000000 | 1.0063934 | 2.0108093 |
| Jul 96 | 07.7500000 | 1.0065642 | 2.0240086 |
| Aug 96 | 07.7500000 | 1.0065642 | 2.0372946 |
| Sept 96 | 07.7500000 | 1.0063525 | 2.0502365 |
| Oct 96 | 07.7661290 | 1.0065779 | 2.0637227 |

| Month/year | Mean basic rate + 2% | Compound interest factor | Cumulative factor |
|---|---|---|---|
| Nov 96 | 08.0000000 | 1.0065574 | 2.0772553 |
| Dec 96 | 08.0000000 | 1.0067760 | 2.0913307 |
| Jan 97 | 08.0000000 | 1.0067945 | 2.1055403 |
| Feb 97 | 08.0000000 | 1.0061370 | 2.1184619 |
| Mar 97 | 08.0000000 | 1.0067945 | 2.1328559 |
| Apr 97 | 08.0000000 | 1.0065753 | 2.1468801 |
| May 97 | 08.2096774 | 1.0069726 | 2.1618495 |
| Jun 97 | 08.4583333 | 1.0069521 | 2.1768788 |
| Jul 97 | 08.6693548 | 1.0073630 | 2.1929071 |
| Aug 97 | 08.9435484 | 1.0075959 | 2.2095642 |
| Sep 97 | 09.0000000 | 1.0073973 | 2.2259089 |
| Oct 97 | 09.0000000 | 1.0076438 | 2.2429234 |
| Nov 97 | 09.2083333 | 1.0075685 | 2.2598990 |
| Dec 97 | 09.2500000 | 1.0078562 | 2.2776531 |
| Jan 98 | 09.2500000 | 1.0078562 | 2.2955467 |
| Feb 98 | 09.2500000 | 1.0070959 | 2.3118357 |
| Mar 98 | 09.2500000 | 1.0078562 | 2.3299978 |
| Apr 98 | 09.2500000 | 1.0076027 | 2.3477122 |
| May 98 | 09.2500000 | 1.0078562 | 2.3661562 |
| Jun 98 | 09.4750000 | 1.0077877 | 2.3845831 |
| Jul 98 | 09.5000000 | 1.0080685 | 2.4038231 |
| Aug 98 | 09.5000000 | 1.0080685 | 2.4232183 |
| Sep 98 | 09.5000000 | 1.0078082 | 2.4421393 |
| Oct 98 | 09.3064516 | 1.0079041 | 2.4614422 |
| Nov 98 | 08.8166667 | 1.0072466 | 2.4792793 |
| Dec 98 | 08.3951613 | 1.0071301 | 2.4969569 |
| Jan 99 | 08.0483871 | 1.0068356 | 2.5140251 |
| Feb 99 | 07.5535714 | 1.0057945 | 2.5285927 |
| Mar 99 | 07.5000000 | 1.0063699 | 2.5446995 |
| Apr 99 | 07.3083333 | 1.0060068 | 2.5599851 |
| May 99 | 07.2500000 | 1.0061575 | 2.5757483 |
| Jun 99 | 07.0750000 | 1.0058151 | 2.5907264 |
| Jul 99 | 07.0000000 | 1.0059452 | 2.6061288 |
| Aug 99 | 07.0000000 | 1.0059452 | 2.6216228 |
| Sep 99 | 07.1833333 | 1.0059041 | 2.6371012 |
| Oct 99 | 07.2500000 | 1.0061575 | 2.6533392 |
| Nov 99 | 07.4750000 | 1.0061438 | 2.6696409 |
| Dec 99 | 07.5000000 | 1.0063699 | 2.6866461 |
| Jan 00 | 07.6532258 | 1.0064822 | 2.7040616 |
| Feb 00 | 07.9224138 | 1.0062773 | 2.7210359 |
| Mar 00 | 08.0000000 | 1.0067760 | 2.7394753 |
| Apr 00 | 08.0000000 | 1.0065574 | 2.7574373 |

*Appendix G*

| Month/year | Mean basic rate + 2% | Compound interest factor | Cumulative factor |
|---|---|---|---|
| May 00 | 08.0000000 | 1.0067760 | 2.7761215 |
| Jun 00 | 08.0000000 | 1.0065574 | 2.7943256 |
| Jul 00 | 08.0000000 | 1.0067760 | 2.8132598 |
| Aug 00 | 08.0000000 | 1.0067760 | 2.8323224 |
| Sep 00 | 08.0000000 | 1.0065574 | 2.8508950 |
| Oct 00 | 08.0000000 | 1.0067760 | 2.8702125 |
| Nov 00 | 08.0000000 | 1.0065574 | 2.8890336 |
| Dec 00 | | | |
| Jan 01 | | | |
| Feb 01 | | | |
| Mar 01 | | | |
| Apr 01 | | | |
| May 01 | | | |
| Jun 01 | | | |
| Jul 01 | | | |
| Aug 01 | | | |
| Sep 01 | | | |
| Oct 01 | | | |
| Nov 01 | | | |
| Dec 01 | | | |
| Jan 02 | | | |
| Feb 02 | | | |
| Mar 02 | | | |
| Apr 02 | | | |

# 4: Use of interest tables

**Table 1: Base rates**

The table records changes in Bank of England base rate from 1990 until the date of publication of this book.

**Table 2: Simple interest at 2% above base rate**

The table lists periods during which various base rates were applicable and derives percentages of the principal sum for use in calculating simple interest during the period. Space is provided for updating to allow for changes in base rates as recorded in Table 1, the factor in the fourth column being:

(period in days) × (base rate + 2%)/(365 or 366)

Where a base rate period straddles the beginning or end of a leap year the factor for the period within each year should be calculated separately in order to allow for the different number of days within each year.

Interest for a given period is calculated by adding together all interest factors within the period, including *pro rata* proportions of the starting and ending periods, multiplying the sum of those factors by the principal sum, and dividing by 100.

*Example*

Simple interest at 2% over base rate on £2.65m for the period from 01.02.99 to 31.05.00

Factor = 0.6136986 × 3/28 + 1.2945205 + 1.2513699 + 1.7452055 + 1.1123288 + 1.4376825 + 0.5928962 + 2.4480874* = 9.9478442
Interest = 9.9478442 × £2,650,000/100 = £263,618

* At the date of publication base rate remained at 6%. Hence the factor for the period from 10.02.00 to 31.05.00 is 112 days × 8/366 = 2.4480874

**Table 3: Interest at 2% above base rate compounded monthly**

The table shows the mean rate of interest plus 2% for each month since January 1990, together with a factor by which the principal sum should be multiplied in order to calculate enhancement for the month. The fourth column shows the product of each month's factor and the preceding cumulative factor. Space is provided for updating information in the second, third and fourth columns, as follows.

*Mean base rate +2%*

Where base rate remains unchanged during the month, the figure is that rate plus 2. Where there have been one or more rate changes, the figure is calculated by adding together the product of each period in days and the rate plus 2% applicable to that period, and dividing by the number of days in the month.

Example: January 2000. Factor =
$(12 \times 7.5 + 19 \times 7.75)/31 = 7.6532258$

*Compound interest factor*

The factor is 1 + (mean base rate + 2%) × (days in month)/(days in year × 100)

Example: June 2000. Factor = $1 + (8.0 \times 30/36600) = 1.0065574$

*Cumulative factor*

The factor is the product of each month's factor and the preceding month's cumulative factor.

Example: June 2000. Cumulative factor = $1.0065574 \times 2.7761215$
$= 2.7943256$

*Compound interest calculation*

The factor by which the principal sum should be multiplied is found by dividing the cumulative factor for the termination date of the interest period by that for the commencement date of that period, interpolating for dates within the starting and ending months. This gives the total of principal sum + interest.

Example: Interest at 2% above base rate compounded monthly on £2.65m for the period from 01.02.99 to 31.05.00.
Cumulative factor for 31.05.00 = 2.7761215
Cumulative factor for 01.02.99 = 2.5140251 + 0.0057945/28 = 2.5142320
Principal sum + interest = £2,650,000 × 2.7761215/2.5142320 = £2,926,031
Hence interest = £276,031

# BIBLIOGRAPHY

## Arbitration

Bernstein, R. (1997) *Handbook of Arbitration Practice*, 2nd edn, Sweet & Maxwell, London.

Harris, B., Planterose, R. & Tecks, J. (2000) *The Arbitration Act 1996*, 2nd edn, Blackwell Science, Oxford.

Mustill, M.J. & Boyd, S.C. (1989) *Commercial Arbitration*, 2nd edn, Butterworths, London.

Rutherford, M. & Sims, J.H.M., (1996) *Arbitration Act 1996: A Practical Guide*, FT Law & Tax, London.

## Construction contract law

Abrahamson, M.W. (1979) *Engineering Law and the ICE Contracts*, 4th edn, Spon, London.

Duncan Wallace, I.N. (1995) *Hudson's Building & Engineering Contracts*, 11th edn, Sweet & Maxwell, London.

May, Sir Alan, Williamson, A. and Uff, J. (eds) (1995) *Keating on Building Contracts*, 6th edn, Sweet & Maxwell, London.

Powell-Smith, V., Stephenson, D.A. & Redmond, J. (1994) *Civil Engineering Claims*, 3rd edn, Blackwell Science, Oxford.

Uff, J. (1996) *Construction Law*, 6th edn, Sweet & Maxwell, London.

## Contract law in general

Furmston, M.P. (1996) *Cheshire & Fifoot and Furmston's Law of Contract*, 13th edn, Butterworths, London.

# TABLE OF CASES

**Abbreviations:**

AC           Appeal Cases
BLR         Building Law Reports
CLY         Current Law Yearbook
Lloyds Rep   Lloyd's List Law Reports
WLR         Weekly Law Reports
CA           Court of Appeal
HL           House of Lords
KB           Law Reports, King's Bench

# INDEX